D0856916

Muscles as Molecular and Metabolic Machines

Peter W. Hochachka

CRC Press
Boca Raton Ann Arbor London Tokyo

Library of Congress Cataloging-in-Publication Data

Hochachka, Peter W.
Muscles as molecular and metabolic machines / by P.W. Hochachka.
p. cm.
Includes bibliographical references and index.
ISBN 0-8493-2468-8
1. Muscles—Metabolism. 2. Muscles—Molecular aspects. 3. Muscle contraction. I. Title.
[DNLM: 1. Muscles—physiology. 2. Neuromuscular Junction—physiology. 3. Synaptic Transmission. WE 500 H685m 1994]
QP321.H668 1994
612.7'4—dc20
DNLM/DLC
for Library of Congress 94-6564
CIP

This book contains information obtained from authentic and highly regarded sources. Reprinted material is quoted with permission, and sources are indicated. A wide variety of references are listed. Reasonable efforts have been made to publish reliable data and information, but the author and the publisher cannot assume responsibility for the validity of all materials or for the consequences of their use.

Direct all inquiries to CRC Press, Inc., 2000 Corporate Blvd., N.W., Boca Raton, Florida 33431.

No claim to original U.S. Government works
International Standard Book Number 0-8493-2468-8
Library of Congress Card Number 94-6564
Printed in the United States of America 2 3 4 5 6 7 8 9 0
Printed on acid-free paper

Preface

For most of the three decades that I have been doing research, I have not been mentally ready to write this book. The motivation for it arose only late in the game, when I realized that the way many of my colleagues and I organized our research, the way we defined the problems and the way we proceeded to analyze and interpret a number of issues regarding how muscles work have been powerfully constrained and restricted by prevailing paradigms. Often the data were forced into conceptual frameworks even if they fitted poorly, or not at all. Dogma often prevailed over data. Being ready to write this book meant that I had to be ready to discard several traditional intellectual trappings that have characterized this field.

The approach I have taken is to explore the paradigm of muscles as molecular and metabolic analogs of man-made machines in which the structures and functions are exquisitely integrated and matched to each other. The analysis begins with a reductionist approach: reviewing from gene to expression the integral machine components—their synthesis, their molecular form, and their molecular physiology. The working parts of the complete muscle machine are proteins—contractile proteins, structural proteins, enzyme proteins, channels, pumps, transporters, and so forth—and a conservative count indicates well over 100 such machine parts in any given muscle. Since most occur as cell-specific isoforms, the random assortment of these machine parts or protein isoforms in theory could generate an astronomical number of muscle machines (fiber types). But this does not occur and the question is, why not?

To attack this problem, I complemented the reductionist approach with an integrationist/adaptational one that again extended from genes to proteins, and that is where the fun really began. Analysis of how these protein components work led to the realization that at all levels of the muscle machine—at information transfer to the contractile elements, at the energy consuming contractile elements per se, at the energy consuming relaxation processes, and at the energy regeneration pathways—the system structure is being driven more and more by highly efficient interactions between machine components, less and less by diffusion-based processes proposed to be dominant in traditional paradigms. The insight of key requirements for highly efficient interactions (usually between interlinked protein parts of the machine) perhaps more so than any factor led to a reevaluation of how muscle function is regulated, and it is here that major paradigm shifts are proposed. In the first place, traditional regulation theory based on feedback loops is downgraded to a secondary or fine-tuning role, while the major control mechanism involved in large-scale change in rate of function is considered to involve the masking or unmasking of the catalytic potentials of proteins in pathways of energy utilization and of energy production. In terms of the classical enzyme velocity equation, $V_{max} = e_o \times k_{cat}$, coarse level control seems to revolve around e_o regulation, whicle fine control seems to be achieved by substrate, product, or modulator-mediated control loops; this seems to apply to essentially all steps in integrated muscle machine function. This paradigm shift is required in order to account for very large changes in rates of muscle function with modest and frequently immeasurable changes in substrate and product concentrations (the classical parameters of feedback and feedforward control loops in traditional theory). Secondly, the only metabolite signal that changes by exactly the right amount (exactly in proportion to change in work) and exactly at the right time is molecular oxygen. However, because oxygen regulation cuts in at concentrations two or more orders of magnitude higher than required for saturation of aerobic metabolism, oxygen regulation of muscle work is postulated to require an oxygen sensing and signal transduction system aimed at numerous protein targets involved in steady-state aerobic muscle work (this perhaps is the most radical paradigm shift called for in the book). Thirdly, in the last part of the book, I argue (i) that these kinds of characteristics are made possible by highly efficient interacting components of the muscle machine and (ii) that the interaction requirements limit (or determine) the number of acceptable combinations of protein isoforms for machine assembly. In fact, the evidence suggests that the more highly specialized the muscle type, the further one moves from the extreme of infinite assortment possibilities and infinite numbers of machine varieties. In super-specialized cases, typically only one fiber type is found, implying that instead of random assortment of isoforms or machine parts, only specific and often unique combinations can work in acceptable fashion.

One of the unexpected spinoffs of writing this book was the recognition, which all scientists probably consider from time to time, of how powerful are the constraining forces of prevailing paradigms in shaping thinking and research in science. When first introduced, new theories expand insight and are intellectually liberating, but the exact opposite can occur, especially in problem areas that are relatively intractable for prolonged time periods (well illustrated in the field of regulation of muscle energetics). In such cases, prevailing paradigms become prevailing dogmas which tend to stifle creativity and to imprison the imaginative mind. For me, the risk of being overly constrained by currently popular concepts has been greatly reduced by a cadre of recent graduate students and postdoctoral fellows whose responsiblities included keeping me awake and on my intellectual toes. For this I express special thanks to Peter Arthur, Jim Ballantyne, Les Buck, Maggie Castellini, Michael Castellini, Geoff Dobson, Chris Doll, Jeff Dunn, Brian Emmett, Chris French, Uli Hoeger, Kathy Keiver, Steve Land, Gord Matheson, Juan Merkt, Tom Mommsen, Chris Moyes, Jennifer Nener, Wade Parkhouse, Trish Schulte, Eric Shoubridge, Carole Stanley, Jim Staples, Raul Suarez, Jean-Michel Weber, and Tim West. My evaluation of how muscles work also was profoundly influenced by several colleagues and collaborators. For this I wish to express special thanks to Hiroke Abe, Peter Allen, Bob Balaban, Marilene Bianconcini, Britton Chance, Kevin Connely, Dick Connett, Eva Danulat, Michael Hogan, Hans Hoppler, Marty Kushmerick, Dick Taylor, Vera Val, Peter Wagner, Ewald Weibel, and Warren Zapol. None of the above, however, should in any way be 'blamed' or be asked to assume any responsibilities for any of my interpretations! Finally, I would like to acknowledge Dr. Rob Stevenson and his colleagues for providing the electron micrograph of insect flight muscle used on the cover of this book.

Most of the intellectual activity required for this book was supported by NSERC (Canada), to whom I am most grateful. Also, in 1993 I was awarded the Science Prize of Canada by the Killam Foundation and the Canada Council. I was greatly honored and proud to receive the prize, which gave me added time and impetus and thus facilitated the completion of this project as well as other work in my laboratory.

Peter W. Hochachka
University of British Columbia
Vancouver, Canada

Dedication

To Claire, Gail, Gareth
who for all these years have tolerated the heart of a wanderer

and to Brenda
who for all these years has tolerated the mind of a dreamer

Abbreviations

ACh	acetylcholine
AChE	acetylcholinesterase
AChR	acetylcholine receptor
ADP	adenosine diphosphate
AMP	adenosine monophosphate
ATP	adenosine triphosphate
APK	arginine phosphokinase
cAMP	cyclic AMP
CA	carbonic anhydrase
CNS	central nervous system
CoA/CoASH	oxidized/reduced coenzyme A
CPK	creatine phosphokinase
Cr	creatine
DNA	deoxyribonucleic acid
DPG	diphosphoglycerate
EC coupling	excitation contraction coupling
ECF	extracellular fluid
e_o	enzyme concentration
ETS	electron transfer system
EPO	erythropoitin
FAD/FADH	oxidized/reduced flavin adenine dinucleotide
FBP	fructose-1,6-bisphosphate
FFA	free fatty acids
FG	fast-twitch glycolytic fibers
FOG	fast-twitch oxidative glycolytic fibers
F6P	fructose-6-phosphate
fSR	free sarcoplasmic reticulum
GAP	glyceraldehyde-3-phosphate
GAPDH	glyceraldehyde-3-phosphate dehydrogenase
G1P/G6P	glucose-1-phosphate/glucose-6-phosphate
GPDH	α-glycerophosphate dehydrogenase
HK	hexokinase
ICF	intracellular fluid
IMP	inosine monophosphate
jSR	junctional sarcoplasmic reticulum
LDH	lactate dehydrogenase
LES	Lambert-Eaton Syndrome
k_{cat}	turnover number per catalytic site
K_{eq}	equilibrium constant
K_m	Michaelis constant
M	myosin
MDH	malate dehydrogenase
MRS	magnetic resonance spectroscopy
MW	molecular weight
mRNA	messenger ribonucleic acid
NAD^+/NADH	oxidized/reduced nicotinamide adenine dinucleotide
NMR	nuclear magnetic resonance
NMRS	nuclear magnetic resonance spectroscopy

PArg	phosphoarginine
PCr	phosphocreatine
PDE	phosphodiesterase
PDH	pyruvate dehydrogenase
PFK	phosphofructokinase
PGI	phosphoglucose isomerase
PGK	phosphoglycerate kinase
PGM	phosphoglucomutase
PK	protein kinase
PKA	protein kinase A
PTK	protein tyrosine kinase
$\dot{Q}O_2$	flow rate
Q_{10}	temperature coefficient (for 10 degree increments)
RBC	red blood cells
RMR	resting metabolic rate
RNA	ribonucleic acid
SO	slow oxidative fibers
SR	sarcoplasmic reticulum
STX	saxotoxin
T system	transverse tubule system
Tn	troponin
TT	transverse tubules
TTI	tension time integral
TTX	tetrodotoxin
Tyr-P	tyrosine phosphate
V_{max}	maximum velocity (maximum catalytic rate)
$\dot{V}O_2$	O_2 consumption rate
$\dot{V}O_{2(max)}$	maximum sustainable O_2 consumption rate

Contents

Chapter 1
FRAMEWORK ... 1
The Setting ... 1
Isoform Machinery for Speeding Up Information Transfer ... 4
Troponin C Isoforms in Speeding Up Contractile Function ... 5
Coadaptation of Energy Demand and Energy Supply Pathways ... 6
Coadaptation and Metabolic Isozymes: The LDH Archetype ... 10
Coadaptation and Emergent Properties ... 11

Chapter 2
NERVE-TO-MUSCLE SIGNALS ... 13
How the Signals Get There ... 13
What Are Channels? ... 14
Acetylcholine: The Signal to Go ... 15
ACh Release Depends Upon Ca^{++} Channels ... 15
ACh-Induced Depolarization Depends on End-Plate Channels ... 16
Structure of End-Plate Channels ... 17
End-Plate Channel Isoforms ... 19
Localization of End-Plate Channels ... 19
Synaptic Transmission Time ... 20
Muscle (and Nerve) Action Potentials Depend on Na^+ and K^+ Channels ... 21
Structure of Na^+ Channels ... 21
Na^+ Channel Isoforms ... 22
Localization of Na^+ Channels ... 24
Isoforms of Delayed Rectifier K^+ Channels ... 24
Excitation-Contraction Coupling ... 25
Ca^{++} Channel Isoforms ... 28
E-C Coupling in Fast- and Slow-Twitch Muscles ... 29
Facilitating Role of Calsequestrin ... 30

Chapter 3
DESIGN OF NERVE-TO-MUSCLE INFORMATION SYSTEMS ... 31
Introduction ... 31
Channels Are Extremely Efficient Catalysts ... 31
Channel Densities Are Usually Rather Low ... 33
Channel Densities May Be Adjusted Upward ... 34
Design Criteria for Presynaptic Signaling Processes ... 35
Design Criteria for Postsynaptic Signal Transduction ... 36
Design Criteria for Na^+ Channel Functions ... 37
Design Criteria for TT and SR Ca^{++} Channels ... 38
Overall Design Principles for Nerve-to-Muscle Information Flow Systems ... 39

Chapter 4
ENERGY DEMAND OF MUSCLE MACHINES ... 41
Introduction ... 41
The Sarcomere: The Basic Contractile Unit ... 42
Sliding Filament Model of Contraction ... 43
Taking Myosin Apart to Identify Functional Domains ... 44

Globular and Filamentous Forms of Actin 45
ATPase Coupling with Filament Movement 45
Troponin and Tropomyosin Mediate Ca^{++} Regulation of Muscle Contraction 47
Adaptable vs. Conservative Aspects of Contractile Components 49
Myosin Isoforms: Patterns and Distribution 49
Actin Isoforms 50
Tropomyosin Isoforms 51
Troponin Isoforms 51
Co-occurrence of Specific Contractile Isoforms 51
Excitation-Contraction (E-C) Coupling: Decoupling of Ca^{++} Channel and Ca^{++} Pump Functions of the SR 52
Ca^{++} Pump Isoform 52
Co-occurrence of Contractile and Regulatory Protein Isoforms 52
Role of Actomyosin ATPase in Adaptation of Muscle Function 53
In Solution, Actomyosin ATPases Are Highly Adapted Catalysts 53
Diffusion Limitation of Enzyme Function Can Be Circumvented 54
The Contractile Cycle as a Channeled Reaction Sequence 55
Evidence For Preferential Access To and From Actomyosin ATPase 57
Minimizing Ca^{++} Diffusion-Based Limits 57

Chapter 5
RETURN TO THE PRECONTRACTION STATE 59
Introduction 59
Ca^{++} ATPase and Sarcoplasmic Reticulum (SR) Structure 59
Ca^{++} ATPase Catalytic Cycle 60
Model of Ca^{++} ATPase Structure 61
Two Ca^{++} ATPase Genes: Two (or More) Ca^{++} ATPase Isoforms 62
Two Ca^{++} ATPases: Differences and Homologies 62
Coadaptation and Design Properties of SR Ca^{++} ATPases 62
Na^+K^+ ATPase and the Na^+ Pump 63
Na^+K^+ ATPase Catalytic Cycle 64
Na^+K^+ ATPase Structure 64
Minimal Design Criteria for Na^+K^+ ATPase as an Ion Pump 65
Na^+ Pump Isoforms Based on α and β Subunit Polymorphism 65
Functional Significance of Na^+ Pump Isoforms 66
Magnitude of the Postexercise Na^+K^+ Imbalance 66
Short-Term Na^+K^+ ATPase Regulatory Mechanisms 66
Long-Term Na^+K^+ ATPase Regulatory Mechanisms 67
Muscle Na^+K^+ ATPase Functional Design Considerations 67

Chapter 6
SUPPLYING MUSCLE MACHINES WITH ENERGY 69
Introduction 69
Three Basic ATP-Synthesizing Pathways in Muscle 69
Utilizing ATP 74
The Nature of Effective Phosphagens 74
Phosphagen Storage 74
Utilizing Phosphagen 74
Phosphagens ''Buffer'' ATP Content 75
Phosphagen End Products 76

Osmotic or Ionic Effects of Phosphagen Mobilization 76
Ancillary Roles of Phosphagens 76
Phosphagens: Their Pros and Cons 77
Anaerobic Glycolysis 77
Glycogen—An Ideal Fermentable Fuel 78
ATP Yields of Anaerobic Pathways 78
Turning on Anaerobic Glycolysis: Hormones and Neurotransmitters 78
Turning on Anaerobic Glycolysis: Enzyme and Isozyme Function 80
The Problem of Anaerobic End Products 82
Upper Glycolytic Limits in Muscle: A Coadaptation Problem 83
Oxidative Metabolism 86
Nature of Endogenous Aerobic Fuels 87
ATP Yields of Aerobic Pathways 87
End Products of Aerobic Metabolism 87
Nature of Exogenous Fuels of Aerobic Muscle Metabolism 89
Storage Sites of Exogenous Fuels 89
Flux Rates Sustainable by Exogenous Fuels 90
Regulating Fluxes of Exogenous Fuels 91
Coordinating Aerobic and Glycolytic Pathways 93

Chapter 7
INTEGRATING ATP SUPPLY AND DEMAND 95
Quantifying Energy Coupling 95
Energy Coupling in Anaerobically-Driven Muscles 96
Human Muscle at Maximum Aerobic Work Rates 97
Setting the ATP Demand: *V* vs *S* or *S* vs *V* 97
Gastrocnemius of the Laboratory Rat 99
Biceps and Soleus of the Laboratory Cat 99
Gracilis and Gastrocnemius of the Laboratory Dog 101
Gastrocnemius of the Laboratory Rabbit 101
Biceps Femoris of the Greyhound—Canine Super-Athlete 102
Leg Muscle of the Thoroughbred—Equine Super-Athlete 102
Calf Muscle of Variably Adapted Humans 103
Serving a Small Muscle Mass with a Large Cardiac Output 103
Flight Muscle of Insects—Champion Animal Athletes 104
Thermally Driven Change in ATP Turnover Rates 105
Pathway Intermediates and the Latent Enzyme Concept 105
Exogenous Control of Energy Coupling 108
Oxygen Sensing in Regulation of ATP Turnover 109
O_2 Sensing—Possible Pathways and Mechanisms 115
Controlling the Physical State of ICF 116

Chapter 8
ISOFORM DEFINITION OF MUSCLE MACHINES 119
Introduction 119
Defining Muscle Fiber Types 119
Nature's Fastest Oxidative Muscles: One Isoform Combination to the Exclusion of All Others? 121
Fish White Muscle: An Anaerobic Type of Muscle Displaying Exceptional Compositional Homogeneity 124

Sound-Producing Muscles: Structurally Homogeneous Muscles Designed for High Frequency, Not Power Output ... 125
Rattler Muscle of the Rattlesnake ... 127
Brain Heater Organ: A Structurally Homogeneous Muscle Designed for Thermogenesis, Not Power Output ... 127
Electroplax: A Structurally Homogeneous Kind of Muscle Designed for Electrical Discharge, Not Power Output ... 130
Why Muscles Specialize into So Few Different Fiber Types ... 130
Glycolytic Function in Chronic Hypobaric Hypoxia ... 132
Role of MDH and LDH in Chronic Hypobaric Hypoxia ... 132
Role of LDH Isozymes in Chronic Hypobaric Hypoxia ... 133
The Fixed Nature of the Lactate Paradox in Andean Natives ... 134
Isoform Basis for Plasticity of Muscle Function ... 134
The Issue of Emergent Properties ... 135

References ... 137

1

Framework

THE SETTING

It is a time of restlessness and fermentation in biomedical science as we approach the end of a millenium. Impressed with the power of our technology and with our skills at manipulating biological systems, today many of us think we stand on the threshold of a brave new world. We can "see" molecules in real time *in vivo* literally in any organ under low or high activity states in health and in disease—living biochemistry in action. We can pull pico-quantitites of prehistoric DNA fossilized in amber, amplify it, make billions of copies of fossil genes (or gene bits), and compare them to homologues living 70,000,000 years and perhaps 0.3 billion generations later—molecular evolution revealed before our very eyes. We can insert or delete, activate or silence specific genes in newly fertilized eggs or even in specific target tissues—chimeric proteins, chimeric cells, chimeric organisms created within our very own laboratories. At a marine mammal conference recently, someone even quipped that their research had assembled and then studied the biochemistry and physiology of bionic seals diving voluntarily in the sea! These are indeed heady times in biomedical science; the potentials seem so enormous and so close to realization (yet the restlessness has extended into our societies, which see the point-counterpoint nature of opportunities and risks arising from modern research).

In the fields of muscle biochemistry and physiology, of exercise sciences, and of sports medicine, most of these visionaries foresee continued enormous strides in our abilities to manipulate muscle properties and performance. That this view is held widely among exercise scientists is not entirely surprising. For at least the bulk of the 20th century, exercise science has monitored a continuous improvement in human performance capabilities—these are steady, across-the-board improvements, largely independent of the type of exercise involved—and there is no indication that the trend is over. Many recent review articles express the recurring attitude in the field: the "Olympic" goal is muscle manipulation using the tools of modern molecular endocrinology, genetics, and pharmacology in order to get more physical work with less fatigue. However, these views of muscle adaptability and of performance enhancement are mostly based on an extremely narrow biological perspective: on research usually with only two species—rats and humans. There is a real possibility of myopia, which is why this may be a particularly opportune time to sit back, have a second look, and reconsider. What is muscle? How does it work? How is it regulated? How is it adapted for specific kinds of biological work? What are its adaptive (especially peak performance) limits? These are the sorts of questions we are to address in this book.

Although all tissues can adapt to varying biological loads and varying environments, muscle is probably the most adaptable tissue in the body. The lay person's view of this adaptability, distilled from the media and from a culture steeped in health and exercise fads, is that muscle properties such as endurance, strength, mass, and speed all are reasonably adaptable. The popular view, at least to a degree, is based on sound evidence; the differences, after all, between weight lifters and marathon runners are dramatic and pretty convincing. Whereas a potential for muscle adaptability is thus generally accepted, it is not so widely appreciated how vast the true biological adaptation ranges and adaptation directions of muscles actually are. It is not widely appreciated, for example, that muscles of homeotherms can be designed to achieve continuous operation at 80 Hz, to complete a relaxation-contraction cycle in 15 ms, and to turn over energy during work at rates some 500 times faster than at rest! Few readers would know about muscles modified for sound production rather than for hard work, where frequency is favored at the cost of power output and where operation at 550 Hz can be sustained indefinitely! Few if any would know about muscles that have been modified into electric organs in which a 2000-fold increase in energy turnover rate is achieved within the 300 ms required for discharge! A ceiling to muscle adaptability might well be assumed by most lay persons and scientists alike, but based on just one species—the human one—the true biological ''height'' of the ceiling is generally grossly underestimated.

That is why one of the main themes running through this book will be the examination of a wide range of muscle systems to gain insights into what can and what cannot change in the structure and function of so-called ''normal'' systems. We will explore the nature of natural biological ceilings in muscles adapted for different functions and purposes in order to distill out the essence of muscle design. We will explore when and how muscle function reaches such ceilings and, in particular, the regulation of the many linked steps in the process. In short, we will review what is currently known of molecular and metabolic design principles of skeletal muscles in order to clarify the rules for and limits of muscle function and adaptability. Whereas much of the insight into these issues will be gleaned from comparative analyses, the arguments used and conclusions derived should be applicable generally, both to animals and man.

A second major theme of this book—one that intertwines with the first—explores the back-and-forth flow of information between reductionist vs. integrative research. The framework for analysis first begins with the assumption that many molecular and metabolic steps contribute to integrated muscle function: muscle activation, excitation-contraction coupling, contraction per se with the concomitant utilization of adenosine triphosphate (ATP), relaxation with its dependence upon ATP utilization, ATP production by various metabolic pathways with their activities being tightly coupled to contraction-relaxation based ATP utilization rates, and so forth. In the end, these disparate functions integrated over time and space lead to muscle work and power output.

Second, it is accepted that many (indeed, most) of the components required for integrated muscle function are proteins, which usually can be expressed in more than one form (so-called isoforms, isoproteins, or isozymes). Isoforms of the components of muscle tissue are formed in a variety of ways: at the genetic level, with different genes specifying the different forms and the timing for their expression; at the translation level, with different splicing patterns generating different isoforms; or at the post-translational level, with modification of the gene polypeptide products generating different isoforms of those products. This means that, in principle, muscle cells could be put together in many different ways—in as many ways as there are combinations of the isoprotein components that together make up the tissue we call muscle.

A question that will arise over and over again in our analysis, therefore, will be whether or not all possible combinations are or ever can be realized. The answer to that question invariably will be negative. Indeed it will be argued that, in some of the most finely tuned and well adapted muscles, a single unique combination from the myriad of statistically possible combinations seems to be selected at the expense of all others. In the extreme, muscles adapted for one specific kind of function are typically very homogeneous, each muscle cell seemingly identical (or nearly identical) to its neighbor, each cell undoubtedly formed of the same unique combination of isoforms as its neighbor, with all other potential combinations being silenced in development and differentiation. This conclusion seems to be valid for muscles that are designed as endurance performance machines and hence must be powered by mitochondrial-based metabolism as well as for muscles that are designed as burst performance machines typically powered by anaerobic glycolysis or by phosphagen hydrolysis. The conclusion likewise seems as valid for muscles designed for very high speed or high frequency performance (which necessarily and simultaneously sacrifice power output) as for muscles that are selected for low frequency function but high power output.

At a molecular and metabolic level of organization, this means that certain, specific isoform combinations are perhaps uniquely suitable for specific kinds of muscle machines because their properties are best matched to each other, presumably because they are the outcome of coadaptation. At the least, our analysis certainly implies that not all combinations are equally effective for any given kind of muscle performance and that, as a result, the ineffective combinations have been rejected during adaptation and evolution.

This interpretation of how muscles are designed is new to biochemistry and molecular biology, where molecular heterogeneity and especially protein isoforms are accepted empirically almost as a kind of accident of nature. However, the coadaptation framework is not new to physiology. In whole-organism physiology, it has long been appreciated in general terms that different links in integrated function necessarily coadapt and thus must be matched to each other; mismatching would result in inefficiency and possibly even disrupted function. This kind of concept was formalized in recent classic studies of Taylor and Weibel (1991) and their collaborators on the path of O_2 from sites of uptake at the lungs to sites of utilization in the mitochondria, during maximum aerobic exercise in a variety of animals. Their studies found that each link in the path of O_2 from uptake sites to utilization sites was reasonably closely matched to all other links; this seemed to be true for sluggish as well as for athletic species, for large animals and for small ones.

A point these workers emphasize is that both structural and physiological adjustments are necessarily involved in achieving matched integrated function. Similar conclusions has been arrived at by Diamond and his coworkers (1991) in their studies of nutrient uptake and the *in vivo* biochemical properties of the vertebrate intestine. Suarez et al. (1990; 1991) are perhaps the first to push this kind of approach effectively into the molecular and metabolic level, showing that the matching required at organ-to-organ integration described by Weibel, Taylor, Diamond and others extends down to the enzymatic and structural properties of energy demand and energy supply pathways of muscles and other tissues. One of the main conclusions of this book—*that the various linked steps in the process of generating specific kinds of muscle work patterns necessarily must be matched one to the other and that not all possible combinations of isoforms are equally suitable for a given kind of muscle machine*—would be viewed by these kinds of physiologists as a special (molecular and metabolic) version of their physiological concepts of symmorphosis and coadaptation and of safety margins in the integration of biological work loads with biochemical capacities. What is missing, however, in the earlier formulations of these concepts is underlying mechanism. Whereas in this book there will be many points at which this same element is still missing, at least in a few cases, the molecular processes accounting for the overt design of integrated properties in fact can be outlined in some detail. Excitation-contraction (EC) coupling in skeletal muscle compared to cardiac muscle is an excellent case in point.

ISOFORM MACHINERY FOR SPEEDING UP INFORMATION TRANSFER

As will be discussed in detail below, the transfer of excitation information arriving at the sarcolemma to the contractile system in the intracellular compartment (i.e., EC coupling) in the heart requires Ca^{++} flux from the extracellular fluid (ECF) through sarcolemmal-positioned, voltage-regulated Ca^{++} channels to the intracellular fluid (ICF) to induce Ca^{++} release. In this muscle, Ca^{++} acts as a cytosolic second messenger, which activates Ca^{++} release and hence muscle contraction in a distinctly diffusion-dependent process. In the higher frequency contractile machinery of skeletal muscle, this diffusional limitation is overcome (or circumvented) by designing an EC coupling system based almost entirely on highly efficient interactions between two protein systems. In this case, the action potential is thought to move down the transverse tubules, which are extremely rich in voltage-sensitive Ca^{++} channels. The voltage-sensing portion of each Ca^{++} channel causes or mediates conformational change, which closes the gap between the transverse tubules, and the junctional sacs of the sarcoplasmic reticulum (jSR); this process allows direct protein–protein interaction with specific protein binding sites on SR Ca^{++} release channels. The interaction is analogous to allosteric activation of enzymes, but in this case, the allosteric effect is to activate (or open) the SR Ca^{++} channels, releasing the signal for muscle contraction. This EC coupling system minimizes diffusion delays and is thus considered more suitable in the faster frequency operation of skeletal muscle than in the slower frequency operation of heart muscle (Catterall, 1991; Ebashi, 1991).

How many and what kind of adaptation processes are needed to convert the slower EC coupling system of the heart to the faster type of EC coupling in skeletal muscle? An insight into an answer to this question arises from studies of myocytes from mice suffering muscular dysgenesis, for an ECF Ca^{++} requirement is developed when one subunit (the $\alpha 1$) of the cardiac Ca^{++} channel is expressed in these diseased skeletal muscle myocytes to restore EC coupling. Furthermore, preparation and expression of chimeric Ca^{++} channels (part skeletal muscle type, part cardiac muscle type) in these diseased myocytes indicate that a specific domain of the $\alpha 1$ subunit of these Ca^{++} channels is sufficient to restore the direct EC coupling system of skeletal muscle; i.e., one segment of one subunit of a specific Ca^{++} channel isoform is seemingly responsible for allowing interaction with the SR Ca^{++} release channel (Catterall, 1991). One segment of one subunit of a specific channel isoform thus approximates one-half of the fundamental underlying molecular mechanism for transforming the EC coupling of a slow muscle into that of a fast one. Although similar details of the other "half" are not yet known, they necessarily involve the SR Ca^{++} release channels, which may also occur in different isoforms that are overexpressed in jSR membranes of skeletal muscle compared to cardiac muscle (Sorrentino and Volpe, 1993; O'Brien et al., 1993). The net effect is that the two interacting components coadapt. *One adjustment is useless without the other, but when they co-occur, there emerges a new property of the interacting system*: a faster allosteric EC coupling mechanism than the diffusion-dependent, second messenger system found in slower muscle cells of the heart. That the two components must be fine-tuned together in coadapted systems such as the skeletal muscle EC coupling mechanism is dramatically illustrated by diseases that alter one or the other of the two interacting Ca^{++} channels (but not both at once). Muscular dysgenesis is a good example already mentioned above in which the voltage-regulated channels are disturbed; malignant hyperthermia is a similarly instructive example of the drastic consequences of SR Ca^{++} release channel dysfunction. No matter how well one of the channels works, if its interacting partner is disrupted, allosteric EC coupling cannot work well or at all. Allosteric EC coupling is a property of the system emerging from the cooperative interactions of its two key component parts.

TROPONIN C ISOFORMS IN SPEEDING UP CONTRACTILE FUNCTION

When we closely examine other steps in integrated muscle function, similar processes become evident, and for the contractile system per se, these can be well illustrated by muscle fiber-specific troponin isoforms. As will be reviewed in different contexts in later chapters, excitation of the sarcolemma and transverse tubules triggers the release of Ca^{++} by the sarcoplasmic reticulum (SR) which then binds to troponin. Troponin (Tn) is a complex molecule composed of three subunit chains, each dedicated to highly subunit-specific functions. TnC binds calcium ions and binds to TnI; TnI binds to actin, and TnT binds to tropomyosin. TnC is a small (18 kDa) acidic protein belonging to a family of proteins sharing a common motif for Ca^{++} binding: a 12-residue loop flanked by two alpha helical segments and termed the helix-loop-helix or EF-hand unit by workers in this field. Together with a residue in a helix near the C-terminal, the loop provides 7 oxygen ligands for Ca^{++} coordination (Grabarek et al., 1992). Although the helix-loop-helix motif is general, TnC occurs in two different isoforms, one specific to fast muscles and the other to slower contracting muscles (slow twitch skeletal muscle fibers and cardiac muscle). The TnC isoform in fast twitch muscles contains two functionally important Ca^{++} binding sites, termed site I and site II, both of which must be filled to trigger contraction. Conformation changes (specifically movement of helix domains B and C away from helix domains A and D) caused by Ca^{++} binding at both sites I and II expose a hydrophobic cavity (the TnI binding site). These Ca^{++}-induced conformational adjustments alter interactions between TnC and TnI; instead of the inhibitory region of TnI binding actin, it now preferentially switches to a binding domain on TnC (Grabarek et al., 1992). In this process, tropomyosin movement is thought to allow actin and myosin to interact and generate contractile force.

In slow twitch muscle, the TnC isoform is missing Ca^{++} site I binding. As a result, slow twitch and cardiac muscles are activated when one, not two, calcium ions are bound per TnC isoform subunit, and the contraction frequency, power output, and strength are all typically down-regulated in these kinds of muscles. Although the full details of the functional consequences of TnC isoform expression are not yet understood, it is clear that at least some characteristics (Ca^{++} kinetics and temperature dependence being classic examples (Harrison and Bers, 1990)) are directly affected by the troponin C isoforms present and by the tropomyosin isoforms with which they interact.

Again it is important to consider how much and what kind of adaptation processes are required to interconvert fiber types. Operationally, for fast to slow twitch type of TnC, this question reduces itself to the kind of adaptation processes that are required to delete the site I Ca^{++} binding site, and our best insights into this issue currently arise from sequence studies and especially from protein engineering (site-directed mutagenesis). These studies show that simple insertions of glycine and valine to replace aspartate residues 1 and 3 in the Ca^{++} site I binding loop change the regulatory properties of TnC. Similar substitutions, including a valine insertion into the Ca^{++} site I binding loop in cardiac or slow muscle TnC isoforms, change the subunit conformation so that acidic residues point out of rather than into the Ca^{++} binding pocket. In this way, relatively minor substitutions in the site I domain completely abolish Ca^{++} binding. In essence, a relatively simple adaptation—a one or two amino acid substitution or insertion—is necessary and sufficient to account for a part of the fundamental underlying mechanism transforming the Ca^{++} binding kinetics of fast muscles to those of slow muscles, and vice versa. As in the EC coupling system discussed above, the details of the other portions of this interacting system are not yet known but necessarily involve tropomyosin and TnI isoforms and their interaction with actin (Grabarek et al., 1992). Since coordinated adjustments in all three systems at once are more effective than simple change only in troponin or only in tropomyosin isoforms, it is not surprising that these systems typically coadapt. One adjustment is relatively useless without the others, but when they co-occur, there emerge new properties of the interacting system: fast vs. slow muscle type of Ca^{++} activation kinetics, fast vs. slow type of thermal responses, and so forth.

COADAPTATION OF ENERGY DEMAND AND ENERGY SUPPLY PATHWAYS

Just as the components of the information transfer system from ECF to ICF and of the contractile system per se are selected for integrated function, so are the components of energy metabolism coordinated with each other and with ATP demanding (or mechanically working) parts of muscle for integrated function. Although many examples of coadaptation of enzymes with each other and with other cellular components will arise in later sections of this book, it is worth illustrating the situation with some particularly clear examples at this point. One of the most intensely researched of such systems is the interaction of specific creatine phosphokinase (CPK) isoforms with cell ATPases to allow preferential access pathways for ATP entering the ATPase binding domain and for products leaving it. The story here originally developed in studies of the interaction between myosin ATPase and CPK in skeletal and cardiac muscles. These studies very convincingly demonstrated that the muscle type (MM) isoform of CPK is bound to the M-line of the sarcomere in large enough amounts (i) to contribute to the electron density of the sarcomere structure and (ii) to constitute a catalytic activity capable of generating ATP at a high enough rate to pace the maximum rate of actin-activated Mg^{++} ATPase activity of the sarcomere (Wallimann et al., 1984). When compared within a species, the amount of M-line-bound CPK MM isozyme and hence its ATP regeneration capacity vary with fiber type; it is up-regulated in the fast twitch type of muscles compared to slow twitch muscles in proportion to similarly higher myosin ATPase capacities in fast twitch muscle. What is more, the binding of creatine phosphokinase to the myofibrillar M-band is isozyme-specific: only MM CPK, and not the BB homodimer (named for its predominance in brain tissue) or MB heterodimer isozymes, can bind in this sarcomere region. Only the MM CPK isozymes can fulfill the roles (i) of matching catalytic capacities of actomyosin ATPase and (ii) of preferentially supplying ATP for the ATPase and thus for contractile function.

Interestingly, the MM CPK association with myosin ATPase is not unique; a similar MM CPK interaction is known for Ca^{++} ATPase (Rossi et al., 1990) and for $Na^{+}K^{+}$ ATPase (Blum et al., 1990). In both cases, the amount of CPK bound is at least equal to or actually exceeds the ATPase capacities of the ion translocating membrane-based pumps.

CPK isozymes are intimately involved not only with ATPases but also with enzymes in glycolysis and mitochondrial oxidative phosphorylation. In the case of the former, much of the MM CPK of skeletal muscles is colocated with enzymes such as muscle aldolase isozymes at the I-band, forming a multienzyme complex bound through weak interactions with actin (Wegmann et al., 1992). There is so much CPK bound at this site that most workers in this field assume the high energy phosphate of ATP formed glycolytically is transferred via PCr to the cell ATPases (Walliman, et al., 1992); i.e., actin-bound CPK at this site serves to link glycolytic ATP production with ATP utilization by cell ATPases (Figure 1–1).

A similar PCr conduit from ATP synthesis sites to ATP utilization sites occurs during aerobic metabolism. Here, a mitochondrial isozyme of CPK unique to skeletal muscle (sarcomeric mCPK) is thought to facilitate formation of contact zones between the outer and inner mitochondrial membranes so as to interact with mitochondrial oxidative phosphorylation. ATP formed at the ATP synthase in the matrix is preferentially directed toward the adenylate translocase and thence to mCPK. PCr formed in this reaction is considered the effective end product of oxidative phosphorylation. It exits through an anion-preferring, high conductance channel (porin) in the outer membrane of the mitochondria and becomes the substrate for ATPase-linked CPK isoforms.

CPK is not the only kinase that binds preferentially at the contact zones between outer and inner mitochondrial membrances; so does hexokinase. The latter enzyme, abbreviated HK, occurs in at least five tissue-specific isozyme forms (Type II predominating in skeletal muscles), and large fractions of the total activity are typically bound to the outer side of the outer mitochondrial membrane at contact zones with the inner membrane. Recent studies have cloned and sequenced hexokinase isoforms from various tissues, including skeletal muscle. The deduced amino acid sequences indicate extensive homologies between different isoforms and different species. The basic hexokinase motif is that of a single polypeptide chain of about 100 kDa molecular weight and a symmetrical structure formed of two halves, the C-terminal half and the N-terminal half. Although both half portions of hexokinase appear to contain glucose, glucose-6-phosphate (G6P), and ATP binding sites, only the C-terminal half of the enzyme displays catalytic activity; despite extensive homology, the N-terminal half of the enzyme is catalytically inert. Current sequence comparisons are consistent with the hypothesis that the enzyme evolved by gene duplication, followed by gene fusion, probably in invertebrate ancestors. Subsequent evolution is thought to have arrived at an enzyme with one half specialized for catalysis (regulated in part at least by product inhibition); the other (the N-terminal) half is thought to have become specialized for different kinds of functions. Included among these are modulation of binding to the contact zones of the outer and inner mitochondrial membranes (Arora et al., 1993). Whereas the entire N-terminal half of the molecule may be involved in regulatory (or at least noncatalytic) functions, residue deletions within the first 15 amino acids of the N-terminus end of HK first suggested a role in HK targeting of the mitochondria. Subsequent studies with chimeric reporter constructs encoding the N-terminal 15 amino acids of HK indicate that this region of the enzyme is necessary and sufficient to confer mitochondrial localization (Gelb et al., 1992). But why should an enzyme whose catalytic purpose is to initiate glucose metabolism be localized to specific mitochondrial binding domains?

For over 15 years, it has been postulated that hexokinase binding to the mitochondria serves two key processes: (i) speeding up catalysis by activating the enzyme

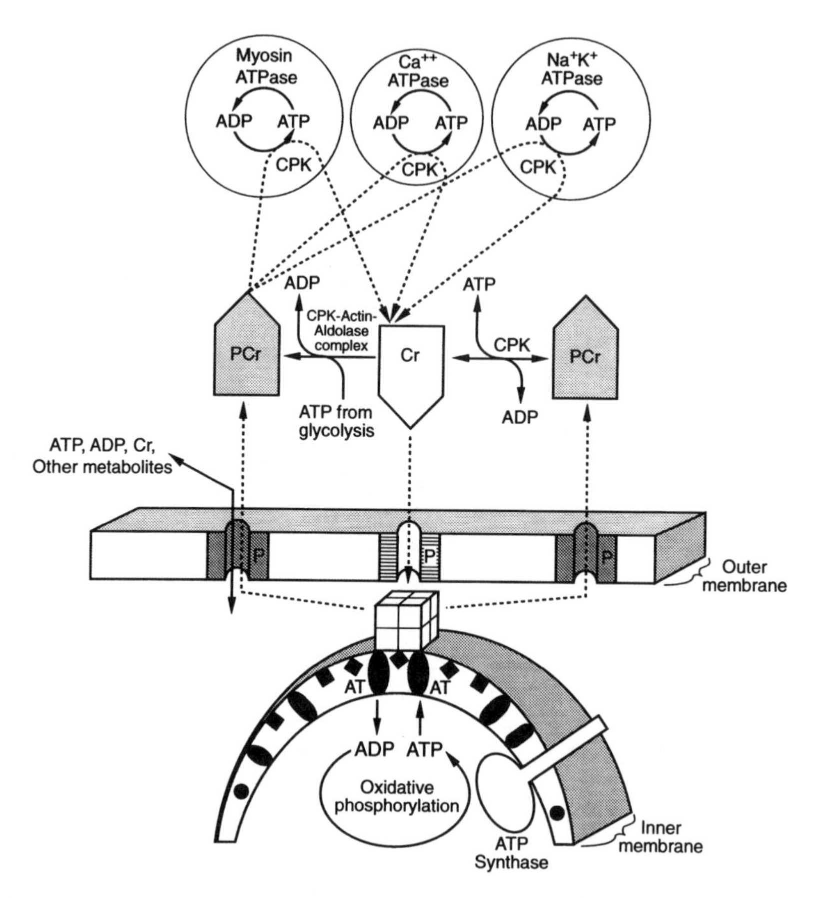

FIGURE 1–1. Postulated roles of creatine phosphokinase (CPK) isozymes in facilitating flow of adenylates between sites of ATP utilization and sites of ATP synthesis in skeletal muscle. Five major compartments for CPK isozymes are indicated: (i) MM CPK associated with glycolytic enzymes and presumed to form PCr from glycolytic ATP, (ii) sarcomeric mCPK localized by binding to porin at contact sites between the outer and inner mitochondrial membranes and presumed to form PCr from ATP formed in oxidative phosphorylation (the product of this reaction, ADP, is thought to have preferential access to the adenylate translocase, AT), (iii) MM CPK associated with actomyosin ATPase, (iv) MM CPK associated with Ca^{++} ATPase, and (v) MM CPK associated with $Na^{+}K^{+}$ ATPase, all three being sites of high and fluctuating rates of ATP utilization. A final site for MM CPK isozyme function (site vi) is assumed to be the cytosol per se, the only site where CPK is considered to play its classical role in maintaining substrates and products in equilibrium. Sarcomeric mCPK is thought to facilitate contact zones actively by binding to a voltage-gated ion selective channel (porin) on the outer membrane and to AT (possibly via an involvement of cardiolipin, represented by dark diamonds) on the inner membrane. Hexokinase is also thought to bind to porin on the outer side of the outer membrane. PCr formed by mCPK is thought to exit via porin channels remote to the contact zone and in a high conductance anionic selective state (shaded); ADP formed by CPK is thought to return via the translocase, while ADP formed by HK and cytosolic Cr flux through a cationic form of porin (hatched) and then are transmitted to AT. See test for other details. Diagram patterned after Wallimann et al., (1992).

(reducing product inhibition) and (ii) supplying a direct (controllable) route for ATP formed at the ATP synthase to the hexokinase for glucose phosphorylation, which would be advantageous during high rates of glucose catabolism. While accepting this overall model of mitochondrial hexokinase function, recent workers (Wicker et al.,

1993) suggest that, because of the relatively high concentration of ATP expected at this site compared to a relatively low [ADP], an additional crucial function of this hexokinase localization is to assure a flow of ADP to the adenylate transolase (for inward flux towards the mitochondrial ATP synthase). This highly attractive hypothesis thus implies a kind of two enzyme (CPK and HK) conduit symmetry around the adenylate translocase: CPK receiving ATP from adenylate translocase and serving as the first main sink for ATP formed in mitochondrial metabolism and hexokinase serving as a kind of catalytic backup reaction, assuring an added constant flux of ADP back to the adenylate translocase to coordinate glucose catabolic rates with oxidative phosphorylation rates.

According to this model of energy demand–energy supply integration in muscle (Figure 1–1), at least six site-specific CPKs carry out unique biochemical functions during accelerated ATP turnover. Of these, only one isoform and one pool of CPK function in the cytosol, presumably operating in solution as a classical catalyst to maintain the reactants at their equilibrium concentration ratios. At each of the other five sites, CPK function appears to be directly dependent upon specific physical interactions with particular target proteins involved in ATP turnover, a situation analogous to those discussed above in EC coupling and in contraction per se (Figure 1–1).

The reason why CPK integration with cellular ATP turnover may have enormous selective value to an organism is perhaps best exposed in studies of CPK coupling with Na^+K^+ ATPase in the electric organ (modified muscle) of electric fish (Blum et al., 1990; 1991). This is because, for practical purposes, the sole source of energy for the Na^+K^+ ATPase in this system is the CPK reaction; anaerobic glycolysis and mitochondrial metabolism are both severely down-regulated and far too sluggish in this modified muscle to satisfy the near-instantaneous and high ATP fluxes required for electrical discharge. Under resting conditions in this kind of system, the catalytic activities of Na^+K^+ ATPase and of CPK (monitored with 31P Nuclear Magnetic Resonance Spectroscopy, or NMRS) are immeasurable; the enormous catalytic activities of the two enzymes are masked, and CPK and Na^+K^+ ATPase in effect are latent enzymes. Upon discharge, the flux through Na^+K^+ ATPase is estimated *to increase by over 2000-fold within several hundred milliseconds (!) and the enormous catalytic capacity of CPK is simultaneously unmasked.* If, in the 300 ms time course of discharge, the catalytic activity of the ATPase were not matched by a similar very large increase in CPK generation of ATP, there would be no way *to avoid local depletion of ATP with local accumulation of products (ADP, Pi, and H^+)* and an equally rapid, drastic inhibition of Na^+K^+ ATPase. However, because of the tight integration of Na^+K^+ ATPase catalysis with CPK catalysis (probably through enzyme–enzyme binding to form a single, functionally linked complex) electrical discharge and discharge of creatine phosphate (PCr) reserves can proceed with minimal perturbation of local and global tissue metabolite and ion (especially ATP and Na^+) concentrations (Blum et al., 1990). In view of similarly large (but not quite so sudden (!)) rest $\rightarrow$ work transitions in flux rates required at Ca^{++} ATPases and at myosin ATPases in vertebrate skeletal and cardiac muscles generally, the same advantages of intimate functional association with CPK would be anticipated and indeed appear to be found (Wallimann et al., 1984). Empirical evidence for this comes from seriously impaired burst work capacities in muscles genetically engineered not to express any MM CPK isoform product (van Deursen et al., 1993).

The direct association of pathway-specific isozymes with other components in integrated muscle function is by no means restricted to CPK interactions with ATPases. Other examples include associations between adenylate translocases and mitochondrial CPK isoforms, glucose transporters and hexokinases, glycolytic enzymes and the chloride-bicarbonate exchanger, Krebs cycle enzymes and inner mitochondrial membrane components, glycogen mobilizing enzyme cascades and glycogen particles, myoglobin and mitochondrial cytochrome oxidase, and so on and so on. This list is by no means complete, but it is sufficient to emphasize that the number of such associations is very large and that the list is growing at a high rate (Pagliaro, 1993) as researchers probe more and more into the inner workings of cell level molecular machines.

How many and what kind of adaptation processes are required for integrating different metabolic enzymes (and indeed, in some cases, different metabolic pathways) with each other and with other macromolecular systems in different types of muscles? The answer to this kind of question will constitute a major part of the chapters that follow, but it may be worth pointing out general answers to it here. Thus, it is evident that at the level of CPK, the main requirement in going from slow to fast twitch muscle is the expression of more MM CPK isozyme so that the CPK catalytic capacity can pace the higher myosin ATPase capacity that is simultaneously expressed. That, too, seems to be the situation in the case of sarcomeric mCPK, where altered expression is known to be regulated by *cis*-acting sequences, including the first intron, the first exon, and an additional 3.36 kb sequence upstream from exon 1 (Klein et al., 1991). The stimuli for turning on this regulatory sequence involve various environmental factors (training, oxygen availability, and so forth). Additionally, sarcomeric mCPK responds to the PCr/Cr pool size. Under conditions of limiting creatine, this mCPK seems to be grossly overexpressed, which leads to mitochondrial abnormalities (Eppenberger-Eberhardt et al., 1991).

In other metabolic pathways, the transition from slow to fast type muscle requires the induction of new isozymes or even new batteries of isozymes; here it is not merely expression of more of the same kinds of enzymes; it is expression of new isoforms to take over the required metabolic steps. For example, the entire pathway of glycolysis is up-regulated in fast twitch muscle compared to slow twitch, and the evidence points to coordinated genetic control of the pathway as a whole. However, in this adaptation, most steps in the pathway are taken over by tissue-specific isozymes. An indication of why this is so is already implied from the above discussion, but this can be even better illustrated with a well-studied case in point, the terminal step in glycolysis, catalyzed by lactate dehydrogenase (LDH).

COADAPTATION AND METABOLIC ISOZYMES: THE LDH ARCHETYPE

LDHs in vertebrate tissues (see Baldwin, 1988) occur minimally as three tetrameric isoforms—usually termed A4 (or M4, because it often predominates in muscle tissue), B4 (or H4, because it often predominates in heart), and C4, which has a relatively tissue-restricted distribution (retina in most species; liver in some fishes). Each LDH subunit is composed of four distinct functional domains: an NH_2-terminal arm (residues 1 through 20), the coenzyme binding domain (the largest domain, including residues 21 through 95 and 118 through 163), the loop domain, which folds over pyruvate when it is bound to the active site (residues 96 through 117), and the substrate binding domain per se (residues 164 through 233). The kinetic properties of LDH holoenzymes are determined by the properties of their constituent subunits. M4 LDHs are generally considered to be pyruvate reductases; they have a high Km for pyruvate and NADH, they are largely refractory to product inhibition, and they are found in tissues with a high reliance on anaerobic glycolysis. H4 LDHs are generally considered to be better lactate oxidases, and these isozymes often are found in tissues with a predominantly aerobic metabolism, often fueled by lactate. Heterotetramers have intermediate kinetic properties, determined solely by their subunit composition. Additionally, M4 and H4 type LDHs display different net charges (and hence different degrees of binding to cellular structures, probably localized to different regions in the muscle cell (Pagliaro, 1993; Murrell et al., 1993)) and different buffering capacities (M4 LDHs have a larger buffering capacity, thought to arise from about two times the content of histidine residues than are present in H4 LDHs).

For isozyme systems like the LDHs, the issue of how many and what kinds of adaptations are required to move from slow to fast type muscles, or vice versa, can be addressed at two levels: (i) regulation of expression and (ii) regulation of function. With regard to the former, muscle-specific glycolytic isozymes are encoded by multiple unlinked genes, which seem to have common *cis*-acting regions in their promoters to bind cellular transcription factors. These regulatory sequences are then thought to allow the unlinked genes to respond coordinately to physiological (or developmental) signals to orchestrate either up- or down-regulation of tissue-specific isozyme versions of the pathway (Webster and Murphy, 1988).

With regard to regulation of catalytic function, most of the major catalytic differences between M4 and H4 isozymes are attributed to several substitutions in the coenzyme binding domain. These substitutions are considered necessary to account for (i) altered affinities for pyruvate and NADH, (ii) altered turnover number, and (iii) altered sensitivity to product inhibition. Altered histidine content and altered net charges of M4 and H4 LDHs (i.e., altered buffering capacity and possibly altered intracellular binding locations) are determined by changes in more than one domain of each subunit (Baldwin, 1988). Hence, *if the LDHs are typical (i.e., are indeed an archetypal isozyme system), then in terms of amounts and kinds of adaptations occurring during evolution of fiber specific and isozyme specific glycolytic pathways, it is fair to conclude that many substitutions must be incorporated at several steps in the pathway from genes to unique protein isoforms to account for the transition from slow to fast kinds of muscle fibers.*

While this conclusion may sound reasonable, is it realistic? Are such isozyme- and cell-specific adjustments relevant? Enzymes such as LDHs usually are considered to operate at or near equilibrium under *in vivo* conditions. Since this is true for both M4 and H4 LDH isoforms, a skeptical reader may well wonder about how much the presence of these kinds of isozymes matters and how we can evaluate or prove their significance in real life. If M4 were deleted from fast twitch muscles, could H4 do just as well? Do their kinetic, buffering, or net charge differences actually matter *in vivo*? The answers to these questions for the human species were answered resoundingly in the affirmative in 1980 when Kanno and his coworkers discovered an M4 LDH-based myopathy distributed through three generations of a Japanese family. Now known to be caused by a 20 base deletion in exon 6 of the M4 LDH gene in skeletal muscle (Mayekawa et al., 1991), this M4 LDH myopathy leads to an almost complete block of anaerobic glycolytic capacity and a remarkable deterioration of muscle performance even at low work rates. Thus, for normal performance of fast twitch muscles, at least in humans, M4 LDH is seemingly indispensible. It is not universally ''indispensible'' since we know that it is almost totally deleted in hummingbird fast twitch flight muscles (The function is taken over by H4 isoform.), but the H4 type LDH here is participating in an entirely different metabolic ''field of play.'' Indeed, in the superfast, extremely oxygen-dependent metabolism of hummingbird flight muscle, the H4 isoform may be as ''indispensible'' as is the M4 LDH in human skeletal muscle (Suarez et al., 1986; 1991).

COADAPTATION AND EMERGENT PROPERTIES

Metabolic isozymes, like the LDHs in metabolic pathway functions of muscle, matter for the same reasons that isoforms matter in EC coupling or in contraction per se, because in each case they form an integral cog in a larger molecular machine. This seems to be true for other tissues as well (Watanabe et al., 1993; Pagliaro, 1993). What is happening (which is exposed through analysis of a few macromolecular components at various working levels in the machinery of muscle) is that three (if not more) processes are occurring at once for improving the transfer of information from the outside to the inside of the cell and for improving the performance of the molecular machinery charged with carrying out mechanical work: one set of processes improves the efficiencies of protein isoform–protein isoform interactions; a second set improves the efficiencies of regulation; the third set minimizes the diffusion-dependence of ion, metabolite, and gas fluxes. In fact, as already mentioned, we shall see that these are not isolated examples and that from several lines of analysis (of information flow pathways to the contractile elements, of the contractile elements per se, of relaxation processes, and of metabolic support pathways) the same conclusion arises again and again: *the system structure is being driven by evolution further and further away from simple diffusion-based, substrate-limited control of fluxes, closer and closer toward fluxes being determined by efficiency of interactions between functionally linked components (often isoform specific components) in the system as a whole.* The integrated properties of these interacting components are considered to be emergent properties in the sense that they cannot be expressed (and often are essentially undetectable) when each component is analyzed in isolation from its normally linked neighbors.

Readers whose training is firmly rooted in traditional biological sciences may well be challenged with a sense of déjà vu. This is because echoes of the concepts (of coadaptation of parts and of emergent properties) expressed here at the molecular and metabolic level of organization are to be found in the classical biological literature on evolution and adaptation, expressed at the whole-organism (often morphological) level of organization. We are quite comfortable with this apparent parallelism and view our new approach and concepts as more mechanistic versions of the older framework.

With this orienting framework in mind, then, let us begin with the information delivery system telling muscle when and how to work.

2

Nerve-to-Muscle Signals

HOW THE SIGNALS GET THERE

We all know, of course, that muscles do not simply work on their own; nerves tell them when and how hard to work. The general path by which the activating signals reach skeletal muscles starts in the central nervous system, but we will pick up the transmission of signals at the motor neurons. Decision for action means the transmission of an action potential down the motor neuron to the presynaptic membranes at the nerve terminals. Transmission from nerve to muscle per se occurs across a gap or a synapse and involves release of a chemical transmitter (acetylcholine, ACh), which must diffuse across the synaptic gap. ACh is then ''received'' at the postsynaptic membrane (it is actually bound to a receptor) in specialized regions of the sarcolemma, termed end plates. Binding of ACh at the end plate leads to depolarization of the end plate membrane, which then spreads as an action potential across the muscle cell membrane. This is the process of excitation. Somehow it leads to an increase in Ca^{++} concentrations in the cytosol, which in turn is widely accepted as the final event triggering muscle contraction. The processes between membrane depolarization and muscle contraction taken together are termed excitation-contraction coupling, an area of very active current research. This much is probably common knowledge to most of our readers.

What may not be so widely appreciated is that each of the steps in the above path of signal from motor neuron to muscle contraction depends upon transmembrane proteins that function as ion channels. In fact, the path for signal transmission from nerve to muscle can be rewritten as follows:

Site and function	**Channel used**
presynaptic membrane: Ca^{++} channel function, release of ACh	plasma membrane isoform of Ca^{++} channel
↓	↓
postsynaptic membrane: ACh receptor (AChR) channel function, depolarization	muscle AChR or end plate channel isoform
↓	↓
muscle membrane: Na^{+} channel and K^{+} channel functions	muscle isoforms of Na^{+} channel and of delayed rectifier K^{+} channel
↓	↓
transverse tubules: Na^{+} channel and Ca^{++} channel function	T. tubule isoforms of Ca^{++} channels
↓	↓
sarcoplasmic reticulum (SR): Ca^{++} channel function	SR isoforms of Ca^{++} release channel

↓ ↓

contraction

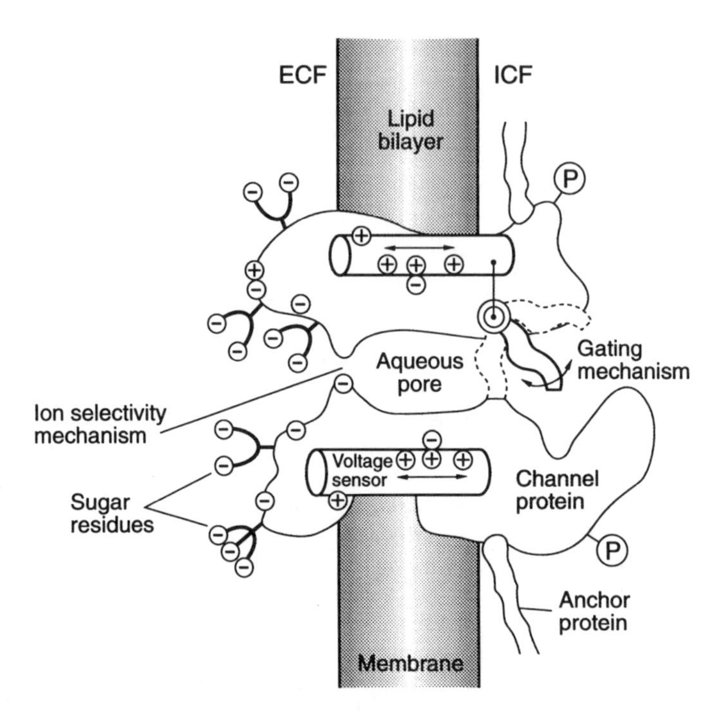

FIGURE 2–1. A hypothetical framework for a voltage sensitive ion selective channel. The channel is diagrammed as a transmembrane macromolecule with a pore through the center. The external surface of the molecule is glycosylated. Key functional regions include an ion selective mechanism, a regulatory or gating mechanism (sensitive to voltage or to ligands), a voltage sensor (whose putative amino acid sequence, deduced from cloning, is highly conserved within homologous channel types, for example, within Na^+ channels), and a phosphorylation site. Modified after Hille et al. (1992).

Thus *the functional participation of at least six channel functions appears to be required in signal transmission from nerve to muscle.* [Furthermore, as we shall see, relaxation (or returning to precontraction states) may require the operation of three or four isoforms of ion-specific, energy-requiring membrane based pumps.] Because these key information processing roles of ion channels may not be widely appreciated, it may be useful to briefly consider what the concept of ion-specific channels is currently understood to be before going on to examine each of the above steps in detail.

WHAT ARE CHANNELS?

A popular view or working model (Hille, 1992) defines a channel as a transmembrane protein embedded in the lipid bilayer of the membrane, anchored in many cases to other membrane proteins or to elements of the intracellular cytoskeleton. The protein is a large oligomer, consisting of several thousand amino acids arranged in one or several polypeptide subunits, with some hundreds of sugar residues covalently linked as oligosaccharide chains to amino acids on the outer face (Figure 2–1).

When open, the channel forms a water-filled pore extending all the way across the membrane. The channel is wider than an ion over most of its length but presumably narrows to atomic dimensions only over a short stretch, the selectivity filter, where the ionic selectivity is determined. Hydrophilic amino acids presumably line the pore wall, and hydrophobic amino acids interface with the lipid bilayer, a so-called amphipathic structure. Regulation of ion flux, or gating, requires a conformational change of the channel that moves a gate into and out of an occluding position. Opening and closing, on the other hand, are controlled by a sensor. In the case of voltage-sensitive channels, the sensor presumably includes many charged groups that move in the membrane electric field during gating. In the case of receptor-regulated channels, the sensor includes a specific binding domain, and receptor binding modifies channel function.

The open-shut nature of gating in single channels can be seen directly with patch-clamp recording methods. For example, single Na^+ channels open briefly to give square pulses of inward current during depolarizing voltage steps. At the single-channel level the gating transitions are stochastic and not exactly predictable. Nevertheless, channels are probably the only proteins where function of a *single* molecule can be monitored precisely for prolonged time periods, and when many records are averaged together, they yield a smoother time course of opening and closing, resembling macroscopic current measurements.

With this model as a framework, let us turn to a detailed examination of the flow of information from motor neuron to muscle.

ACETYLCHOLINE: THE SIGNAL TO GO

For over three decades, we have known that acetylcholine transmits the message at the neuromuscular junction. Arrival of the action potential at the presynaptic motor nerve terminals leads to release of acetylcholine (ACh) in previously prepackaged vesicles. Released ACh molecules diffuse across the synapse and are bound to ACh receptors (AChR) on the postsynaptic membrane, a process that initiates depolarization of the sarcolemmal membrane. The process is terminated by release of ACh and its rapid hydrolysis by acetylcholinesterase. Both the release of ACh (from the presynatic terminals) and ACh-dependent depolarization of postsynaptic membranes depend upon the controlled opening and closing of transmembrane proteins that function as channels or pores, but with different ion specificities.

ACh RELEASE DEPENDS UPON Ca^{++} CHANNELS

As at all chemical synapses, depolarization of the motor nerve terminal opens presynaptic voltage-dependent Ca^{++} channels, permitting external Ca^{++} ions to enter and trigger the exocytosis of prepackaged vesicles of transmitter. In overall outline, we now know how this occurs.

In the resting presynaptic cell, the cytoplasmic free calcium levels are held very low by the combined actions of a Na^+-Ca^{++} exchange system on the surface membrane and ATP-dependent pumps on mitochondria and other intracellular organelles, such as the sarcoplasmic reticulum. The normal resting $[Ca^{++}]$ is so low that it is difficult to measure, but it probably lies in the range 10 to 300 n*M* in living cells. When the membrane potential falls far enough, Ca^{++} channels in the surface membrane open, and Ca^{++} ions enter the cytoplasm, raising the local $[Ca^{++}]$ transiently until the buffering and pumping mechanisms tie up or remove the extra Ca^{++}. In contrast to the situation with Na^+ or K^+ ions, the normal $[Ca^{++}]$ is so low that it is easily increased 20-fold during a single depolarizing response in a cell with Ca^{++} channels. This increase triggers a two-step ACh release process. First, prepacked

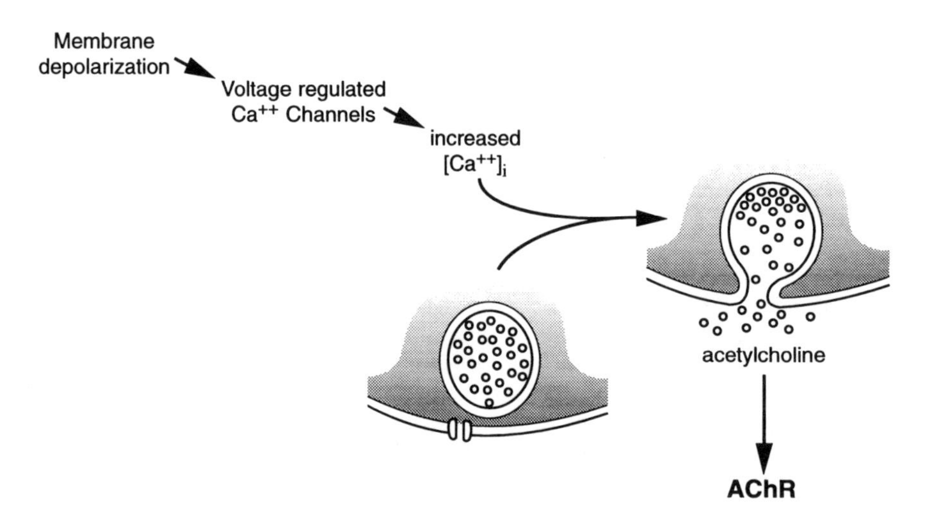

FIGURE 2–2. Sequence of events in Ca^{++} control of secretion of acetylcholine. The binding of vesicles filled with acetylcholine to the cell surface membrane is mediated by some currently unknown fusion-inducing molecules. Acetylcholine is released by exocytosis of a vesicle and diffuses to the acetylcholine receptor (AChR) on the postsynaptic membrane. Modified from Hille et al. (1992).

vesicles of secretory products associate with the cell surface membrane in conjunction with some unknown, Ca^{++}-sensitive, fusion-inducing molecules. Secretory signals cause Ca^{++} ions to enter through Ca^{++} channels on the plasma membrane or to be released from intracellular stores. Second, secretory product (ACh) is released by exocytosis of vesicles after Ca^{++} ions have triggered the necessary membrane fusion (Figure 2–2).

Early quantitative experiments with frog neuromuscular junction indicated that the probability of release of ACh vesicles during an action potential increases as the fourth power of $[Ca^{++}]$ in the external medium. The steep $[Ca^{++}]$ dependence was first explained by Katz and Miledi's (1967) proposal that the presynaptic action potential opens Ca^{++} channels in the presynaptic terminal, letting in a pulse of Ca^{++} ions, which in turn react with an intracellular Ca^{++} receptor. Several such receptors cooperatively control the release of one vesicle from the terminal, thus contributing to the fourth power $[Ca^{++}]$ dependency. The nature of the Ca^{++} receptor for secretion of neurotransmitter is not known, but calmodulin seems a good candidate in some secretory processes (Means et al., 1982).

Processes regulated by $[Ca^{++}]$, such as ACh release, thus acquire a secondary voltage dependence from the voltage dependence of opening of neural plasma membrane-specific isoforms of Ca^{++} channels (Rosenberg et al., 1986; also, see below for further discussion). Functionally, of course, this is crucial and may be viewed as a key part of the transduction of activating signal from presynaptic motor nerve to muscle.

ACh-INDUCED DEPOLARIZATION DEPENDS ON END-PLATE CHANNELS

Just as the release of ACh from preloaded vesicles depends on Ca^{++} channel function, ACh-dependent depolarization of the postsynaptic membrane depends upon

ACh-activated channels found in the neuromuscular postsynaptic membrane. This organization is common to all vertebrate skeletal muscles. If two ACh molecules bind to receptor sites on the end-plate channel protein, a wide pore, permeable to several cations, opens and initiates the depolarization. In frog sartorius at 22° C, end-plate channels open for only about 1 ms in response to the binding of two molecules of ACh; in mammalian muscles the open time is perhaps a millisecond. When they close, the agonist (ACh) can leave the receptor to be hydrolyzed by acetylcholinesterase. During a normal synaptic transmission, the cleft concentration of ACh falls so rapidly that the channel is not likely to be activated a second time by rebinding of transmitter. For this apparently simple job, the microscopic gating properties are remarkably complex, including multiple openings separated by tiny closed periods within the 1 ms dominant time constant, a little voltage dependence, and a variety of functional states (three or more). Each of these subtle microscopic properties might possibly confer an important adaptive advantage. Alternatively, each might be a biologically unimportant consequence of the major opening mechanism. In any case, they indicate that the microscopic regulatory properties of even apparently simple channels can be highly complex (Wan and Lindstrom, 1984).

STRUCTURE OF END-PLATE CHANNELS

Like all channels, the end-plate channel is a glycoprotein, however, more is known of its structure than of any other channel. Before the mid-1960s there were no serious thoughts about the chemical structure of channels in general, and there was certainly no focus on proteins as possible channels. However, by 1973, both the end-plate channel and the Na^+ channel were identified as proteins, and their chemical purification was under way. This progress depended on technical developments in protein chemistry and in pharmacology. Methods were appearing for solubilizing and purifying membrane proteins without destroying their function, and selective toxins were being found that bind to the channels with high affinity and that could be used in radioactive form to identify channel molecules during purification. Initial progress gained great impetus from a rich source of end-plate channels: the electric organ of *Torpedo,* the electic ray. This muscle-derived organ, designed to deliver a high-current shock to prey, is a battery made from stacks of hundreds of cells in series. Each generates a pulse of current through a vast array of AChR channels in response to impulses in a presynaptic cholinergic axon. One whole side of each cell is, in effect, a giant end plate. Another good source of the AChR channel is the electric organ of the electric eel *Electrophorus electricus,* a teleost fish. Isolation and purification procedures typically use ^{125}I-α-bungarotoxin as an assay for the end-plate channel.

In preparations from both species (Figure 2–3), the end-plate channel occurs as a 250 kDa oligomeric protein, formed from four different kinds of subunits with molecular masses of 40, 50, 60, and 65 kDa (Wan and Lindstrom, 1984). The peptides α, β, γ, and δ exist in a pentameric stoichiometry, $\alpha_2\beta\gamma\delta$ in the original complex, making a total molecular mass of 268 kDa. In addition, about 75 carbohydrate residues (galactose, mannose, glucose, and *N*-acetylglucosamine) are attached as oligosaccharide chains, some to each peptide subunit, and the protein has several sites of phosphorylation. As in all other membrane proteins, glycosylation probably defines regions of protein facing the extracellular medium. The majority of the receptor is exposed on the extracellular surface and four or five transmembranous-helical domains are thought to exist in each subunit. The cation channel seems to traverse the membrane through the center of the receptor, and the subunits are apparently oriented

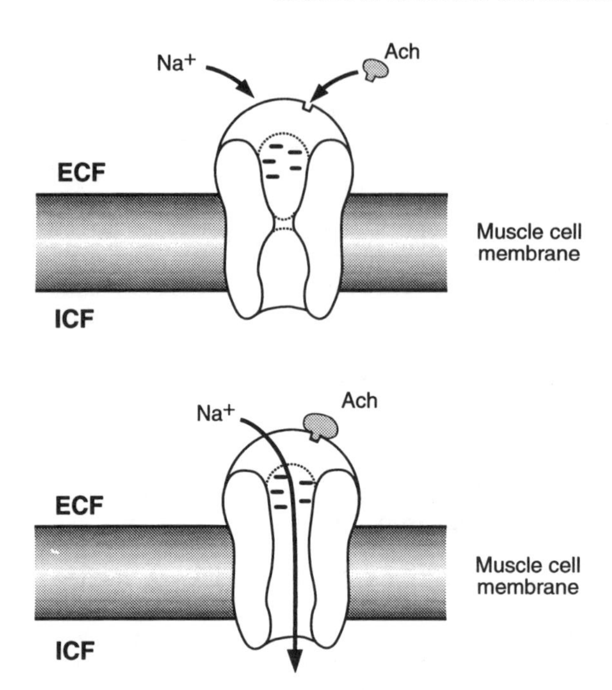

FIGURE 2–3. Diagrammatic model of the acetylcholine receptor (AChR) in the closed state (upper panel) and open state (lower panel), induced by conformational adjustments that occur when acetylcholine (Ach) is bound.

around this central channel. Each subunit donates a domain to form the channel, as might the staves of a barrel. Viewed from the top by electron microscopy, the receptor looks like a doughnut; viewed from the side, it looks somewhat like a funnel (Wan and Lindstrom, 1984). Experiments with antibodies and water-soluble covalent modifiers show that each subunit also displays a cytosolic domain exposed to the intracellular medium. Hence α, β, γ, and δ each extends fully across the membrane. *The intact, purified AChR complex includes all the major functions of the ionic channel since, when it is incorporated into lipid bilayers, it assembles into channels that express appropriate ionic selectivity, conductance, and responses to agonists and antagonists* (Tank et al., 1983).

Knowledge of the partial amino acid sequences made it possible to clone and sequence DNA copies of the messenger RNAs for the entire end-plate channel (Noda et al., 1983b). Messenger RNAs from *Torpedo* organ were copied by the enzyme reverse transcriptase to yield "cDNA" transcripts. The transcripts were then inserted into the DNA of a plasmid used to transform *Escherichia coli* cells, which were grown up as clones containing random samples of the original *Torpedo* electric messenger sequences. Hundreds of thousands of clones were screened to find some that matched sequences of a single AChR subunit. Each selected plasmid DNA could then be sequenced, and its triplet codes could be read off to give the primary amino acid sequence of an entire AChR subunit. As expected from earlier protein chemistry studies, the sequences showed extensive homology. Similar work with the electric eel and mammalian muscle AChR reveals that they too comprise an $\alpha_2\beta\gamma\delta$ complex having subunits with sequence homology to those of the elasmobranch (e.g., Noda et al., 1983a). More recently, up to seven genes for the α subunit and four for the β subunit have been sequenced (Hille, 1992).

END-PLATE CHANNEL ISOFORMS

Although homologous amino acid sequences are instructive in understanding conservative features of channel structure, an equally important insight arising from these studies is that *end-plate channels occur in different forms* often displaying some tissue specificity. Earlier studies had already clearly established that cardiac muscle AChR channel differed strikingly from the skeletal muscle isoform; the latter displays nicotinic pharmacology, while the former is specifically reactive to muscarine derivatives. [Muscarinic AChR channel isoforms are found in the atrial ventricular and sinu-atrial nodes of the heart, in smooth muscle, in ciliary eye muscles, and in parts of the CNS, and display very different regulatory properties (see Mattera et al., 1986)]. More recent cloning studies suggest that while the AChR channels of both nerve and skeletal muscle display nicotinic pharmacology, their α subunits display a number of key amino acid sequence differences; they therefore can be classified as true, tissue-specific isoforms (Boulter et al., 1986). Even the end-plate channel proteins of fast and slow skeletal muscle types display significant functional differences (Dionne, 1981). Although these could be based upon underlying structural (primary-sequence) differences formed either as products of specific genes or by post-translational modifications, more recently it has been possible to isolate and sequence at least seven genes for the α subunit and four for the β subunit (Hille, 1992). The functional implication of all these kinds of studies is the same, *namely, that AChR channel isoforms are the molecular basis for at least some of the membrane properties that are unique to various tissues and organs* in the body. Included among these features are specific membrane localization patterns.

LOCALIZATION OF END-PLATE CHANNELS

In muscle, end-plate channels, with their built-in ACh receptors, are densely clustered on junctional folds of the surface membrane in the end-plate region, immediately opposite the unmyelinated, presynaptic nerve terminal (Figure 2–4). There are about 1×10^7 receptors per end plate, one end plate per fiber, and less than 40 μg of channel protein per kilogram of muscle (Lindstrom, 1985). AChR channels open in response to nerve-released transmitter and depolarize the neighboring end-plate area. Normally, this depolarization, the end-plate potential (epp), is large enough to excite a propagated action potential and a twitch in the muscle. However, it can be reduced experimentally to a subthreshold depolarization if the ACh receptors are partially blocked by a low concentration of a competitive receptor blocker like the alkaloid, curare, or by a practically irreversible blocker, such as the snake neurotoxin, bungarotoxin. The tight binding of snake neurotoxins makes them excellent tools to label, count, or extract end-plate channels. Autoradiography with $[^{125}I]$-α-bungarotoxin shows a dense packing of almost 20,000 binding sites (or 10,000 channels) per square micrometer in the top of the junctional folds of the postsynaptic membrane that lie opposite the active zones of the nerve terminal (Fertuck and Salpeter, 1976; Matthews-Bellinger and Salpeter, 1978). These densities are about one-third the theoretical maximum (at which the channel proteins would be close-packed like bricks in the sarcolemma, theoretically making an essentially lipid-free membrane) and are higher than observed for any other channel (Hille, 1984).

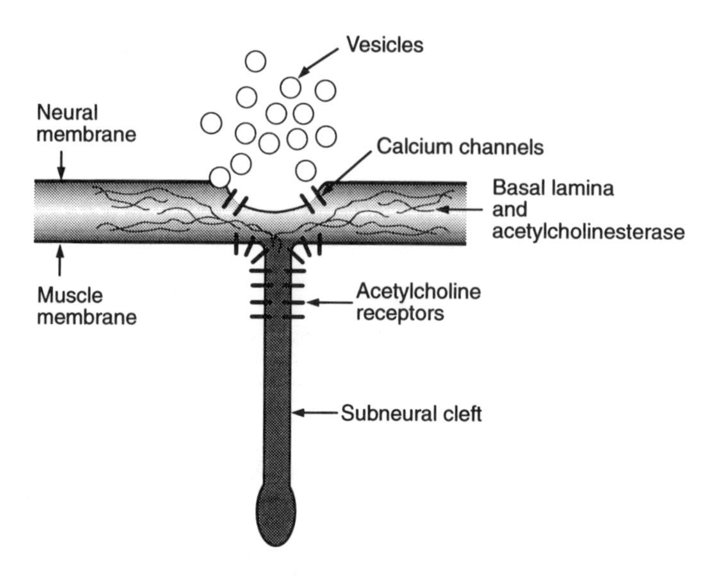

FIGURE 2–4. Organization of the neuromuscular junction, emphasizing the colocalization of release sites on the neural membrane, with high density clusters of acetylcholine receptors at the mouths of subneural clefts.

Such specific localization is not inherent to end-plate regions. In embryonic muscle, uninnervated muscle, or denervated muscle, end-plate channels appear randomly distributed throughout the sarcolemma. Upon differentiation and innervation, already-synthesized end-plate channels are specifically localized to neuromuscular junctions, where they are ultimately needed. It is possible, although not yet proven, that one function of AChR isoforms may be to facilitate proper localization during tissue differentiation. Since the signal inducing proper localization is nerve-derived, it is possible that isoforms facilitate the process through inducer-specific binding sites. Another anchoring mechanism may involve interactions of the cytosolic domain of the channel with cytoskeletal elements, a process that may also select for tissue-specific channel isoforms (Wan and Lindstrom, 1984).

SYNAPTIC TRANSMISSION TIME

In considering the time required for signal transmission from motor nerve to muscle, we will ignore any delays due to conduction time following nerve stimulation in the first place. On the presynaptic side, electrophysiological and morphological experiments show that presynaptic Ca^{++} entry controls secretion by increasing the probability of all-or-nothing exocytosis of vesicles of ACh. In normal circumstances, enough Ca^{++} enters the presynaptic terminal to release an average of 100 to 300 quanta (vesicles) of ACh per impulse within a fraction of a millisecond. For example, the delay between presynaptic depolarization and Ca^{++}-dependent transmitter vesicle release is < 200 μs in fast mammalian synapses at 37° C. On the postsynaptic side, as noted previously, the dominant time constant associated with end-plate channel gating (i.e., with the decay of end-plate current) approaches 1 ms in duration (Hille, 1984).

Following this delay, end-plate depolarization initiates an action potential which is transmitted along the muscle cell.

MUSCLE (AND NERVE) ACTION POTENTIALS DEPEND ON Na^+ AND K^+ CHANNELS

For over 30 years now, we have known that the action potential involves transient changes in membrane permeability to Na^+ and K^+ ions. The sodium permeability of the membrane rises rapidly and then decays during depolarization. In earlier work (e.g., Hodgkin and Huxley, 1952a,b), Na^+ conductance was said to be activated and then inactivated. In newer terminology, we would say that Na^+ channels activate and then inactivate. Much recent research has focused on untangling the processes of activation and inactivation.

In the Hodgkin-Huxley analysis, activation is the rapid process that opens Na^+ channels during a depolarization. Inactivation is a slower process that closes Na^+ channels during a depolarization. Once Na^+ channels have been inactivated, the membrane must be repolarized or hyperpolarized, often for many milliseconds, to remove the inactivation. Inactivated channels cannot be activated to the conducting state until their inactivation is removed. The inactivation process overrides the tendency of the activation process to open channels. Thus, inactivation is distinguished from activation in its kinetics, which are slower, and in its effect, which is to close rather than to open channels during a depolarization. Inactivation of Na^+ channels accounts for the loss of excitability that occurs if the resting potential of a cell falls by as little as 10 or 15 mV—for example, when there is an elevated extracellular concentration of K^+ ions or after prolonged anoxia or metabolic block.

On depolarization, with only a minor time lag, the K^+ conductance also increases, in this case, along an S-shaped time course; on repolarization, the decrease in K^+ conductance is exponential. These changes in K^+ conductance also depend upon the operation of a class of K^+ channels called the delayed rectifiers. Only millisecond time periods are required for the opening and closing of Na^+ and K^+ channels during a propagated action potential. After local circuit currents begin to depolarize the membrane, Na^+ channels activate rapidly, and the depolarization becomes regenerative, but even before the peak of the action potential, inactivation takes over, and the Na^+ permeability falls. In the meantime, the strong depolarization slowly activates K^+ channels, which in large measure are responsible for the outward current needed to repolarize the membrane. These voltage-dependent regulatory (or gating) properties almost certainly depend on channel structure.

STRUCTURE OF Na^+ CHANNELS

As in the case of AChR channels, structural studies of the Na^+ channel could begin only when highly channel-specific toxins (such as TTX and STX) became available in radioactive form as tags, which could be used to follow the progress of the channel protein during purification. Now the Na^+ channel has been purified from *Electrophorus* electric organ, rat skeletal muscle, and rat brain synaptosomes. It is known that when solubilized, the Na^+ channel protein is about 316 kDa in size and consists of three subunits. All Na^+ channel protein preparations contain an unusually large glycopeptide, which carries a TTX binding site. As much as 30% of its mass is carbohydrate, mostly *N*-acetylglucosamine and negatively charged sialic acid (see Catterall, 1984). This means that the peptide has nearly 2000 amino acid residues and 500 covalently attached sugar residues in the form of oligosaccharide chains. Although β-1 and β-2 subunits are present, most of the functional properties reside in the large α subunit, as the cDNA for the α subunit message suffices to express most of the channel properties in oocytes. Most cloning work has focused on the α

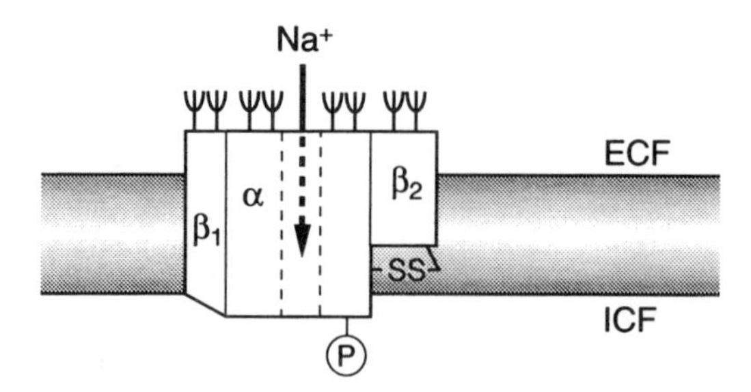

FIGURE 2–5. Na$^+$ channel subunit composition as it is currently understood. Glycosylation (ψ), phosphorylation (P), and intersubunit linkage through disulfide bonds (-SS-) is known from biochemical information; arrangement of subunits is hypothetical at this point. Modified from Catterall (1988).

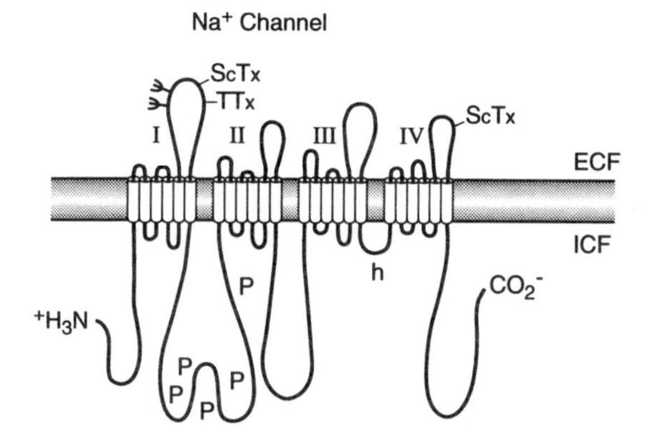

FIGURE 2–6. Postulated transmembrane looping of the α subunit of the voltage gated Na$^+$ channel. Internal repeat units are identified as I, II, III, and IV. Regions identified include glycosylation sites (ψ), phosphorylation (P), scorpion toxin (ScTx) binding sites, tetrodotoxin (TTx) binding sites, and the postulated inactivation site (h). Modified from Catterall (1988).

subunit, for which six different mammalian genes have been fully sequenced (Hille, 1992).

A repeatedly observed feature in all Na$^+$ channel proteins studied to date is the occurrence of four hydrophobic domains, displaying a high degree of sequence homology with similar domains in Ca^{++} channels. Each of these internal repeat units also contains a distinctive segment, S4, with positive charged amino acids at every third residue. As in the latter, the hydrophobic internal repeats are assumed to be sections of Na$^+$ channels that span cell membranes (Figures 2–5 and 2–6), while the S4 is the putative voltage sensor (Catterall, 1986). Transmembrane Na$^+$ flux is presumably mediated by a hydrophilic pore containing a selective ion coordination site, but how the Na$^+$-preferring pore may be formed from constituent subunits is not currently understood.

Na$^+$ CHANNEL ISOFORMS

It is generally held that features such as protein size, oligomeric structure, pore size, pore structure, transmembrane domains, ion coordination site, and neurotoxin binding sites, are similar in Na^+ channel proteins in all cell membranes, from different tissues and from different species; that is, *like channel proteins in general, Na^+ channels appear to be highly conservative.* Nevertheless, three lines of evidence indicate that the Na^+ channel occurs in tissue-specific isoforms analogous to isozymes in the fields of enzymology and metabolism. The first indications of such isoforms are to be found in their differential responses to various toxins and pharmacological agents.

Vertebrate cardiac Na^+ channels, for example, are much less sensitive to TTX than are vertebrate skeletal muscle or nerve Na^+ channels. For example, binding and blocking experiments give inhibitory dissociation constants of 0.5 to 10 n*M* for TTX and STX acting on axons and skeletal muscle of fish, amphibians, and mammals (Richtie and Rogart, 1977), while these values are as high as 1.0 to 6.0 μ*M* for Purkinje fibers and fibers of mammalian heart (Cohen et al., 1981; Brown et al., 1981a). In addition, the Na^+ channels of embryonic neurons and skeletal muscle express TTX insensitivity during development (Spitzer, 1979). These data are consistent with multiple forms of Na^+ channels in each organism, which can be recognized by their different pharmacology. (It is instructive that Na^+ channels of puffer fish and salamanders that make TTX for self-defense are inherently very resitant to the toxin.)

A second line of evidence for Na^+ channel isoforms comes from protein purification studies; as noted above, these studies indicate that the Na^+ channel typically found in nervous tissue consists of three types of subunits (Catterall, 1984), while the muscle Na^+ channel isoform appears to be formed of only two types of subunits (Catterall, 1986).

A third line of evidence for Na^+ channel isoforms comes from studies of tissue differentiation, which indicate different genes for the Na^+ channel are turned on at different stages in the process (Spitzer, 1979).

A fourth line of evidence is based on monoclonal antibodies. By this criterion, Na^+ channel isoforms in central axons vs. peripheral axons, in nerve vs. muscle, and even in transverse tubules vs. the plasma membrane can be distinguished readily.

Finally, molecular biology supplies the most recent line of evidence on Na^+ channel isoforms: cloning studies so far have confirmed six Na^+ channel transcripts from rodents, four from brain, one from skeletal muscle, and one from heart (Hille, 1992).

An important question concerns the functional significance of Na^+ channel isoforms. What functional importance, if any, do they have? Three possibilities suggest themselves. First, the isoforms could serve to impart tissue-specific voltage dependence of the Na^+ channel opening. This is an attractive possibility, which may not be ruled out fully until the molecular basis for voltage dependence is clarified. Recent structural studies, however, indicate 100% sequence homology of a strongly positively-charged peptide found in different Na^+ channels. This is considered to be a part of one transmembrane section of a channel subunit and raises the possibility of a positively charged voltage sensor built directly into the structure of the channel. If this sliding helix model of voltage-dependent gating of the Na^+ channel turns out to be correct, it would appear to be universal (the same in all isoforms). Thus, a role for isoforms in fine tuning voltage sensing is tentatively considered unlikely.

A second possibility is that Na^+ channel isoforms supply a part of the molecular basis for tissue-specific properties of action potentials. Whereas this cannot be ruled out as a possibility, Hille (1984), among others, is more impressed with the *similarity* of functional properties of different Na^+ channel isoforms than he is with their distinctions. A comparison of the time courses of Na^+ conductance changes in nerve and muscle cells from four different phyla indicates that, after correction for temperature, Na^+ channel activation-inactivation time constants do not differ more than two-fold. Thus, we tentatively also rule out the second possibility as an explanation for why Na^+ channel isoforms were developed in the first place.

A third possibility is that Na^+ channel isoforms supply important information for sorting and localization, as suggested in the case of end-plate channels.

LOCALIZATION OF Na^+ CHANNELS

The question of Na^+ channel localization is inseparable from the question of lateral mobility of membrane proteins in general. This mobility, where and when it occurs, is viewed as similar to protein "icebergs" floating in a sea of membrane lipid (Singer and Nicholson, 1972) and is consistent with measurements showing that membrane lipids are fluid. They rotate, their fatty acid chains are mobile, and they change their lipid neighbors over 10^6 times per second. Despite this evidently dynamic lipid sea in which membrane proteins find themselves, many of them, such as Na^+ and end-plate channels of nerve and muscle, are not free to move laterally. In the case of end-plate channels, lateral diffusion coefficients are so low they fall below the limits of measurement, $< 10^{-12}$ cm^2/s. In adult muscle, this immobility helps to preserve differentiated end plates, with α-bungarotoxin binding-site densities of 20,000 per μm^2 in contrast to *nearby extensive extrajunctional membrane, where site densities are as low as 6 to 22 per* μm^2 (Hille, 1984).

In vertebrate skeletal muscle, direct mobility measurements of Na^+ channels similarly reveal no lateral movement and a diffusion coefficient of $< 10^{-12}$ cm^2/s (Stuhmer and Almers, 1982). This channel immobility in membranes is consistent with highly specific distribution of Na^+ channels (high channel densities near motor end plates and along transverse tubules; channel densities decreasing with distance from end plates and decreasing near tendons) in both fast and slow type muscles. A similar patchy or nonrandom distribution of delayed rectifier K^+ channels is also observed in skeletal muscle (Almers et al., 1983). In nervous tissue, within-membrane immobility is consistent with the focal concentration of thousands of Na^+ and K^+ channels per square micrometer in the nodal membrane of myelinated axons with almost none in the paranodal membrane only 1 μm away.

What is the basis or mechanism underlying membrane protein immobility? Unfortunately we do not know. However, *even if the* Na^+ *channel properties required for within-membrane anchoring are not known, it is obvious that these requirements may differ drastically between tissues;* conditions for anchorage in sarcolemma, for example, undoubtedly differ from those in axonal nodes. We consider that satisfying such differing localization requirements is a plausible (perhaps *the* most plausible) biological function of tissue-specific Na^+ channel isoforms, which helps to explain why, in the absence of large kinetic differences, these isoforms arose in the first place.

ISOFORMS OF DELAYED RECTIFIER K^+ CHANNELS

The K^+ channels associated with generating action potentials also vary in their properties within tissues of the same organism (see Miller, 1991). For example, delayed rectifier K^+ channels of frog skeletal muscle require 8 m*M* external TEA for half blockage, compared to the 0.4 m*M* needed at the node. The gating kinetics of these channels differ as well. We now know that K^+ conductance of many cells not only activates with depolarization, but it also inactivates, a phenomenon overlooked in earlier work. In frog skeletal muscle, K^+ channels inactivate exponentially with a

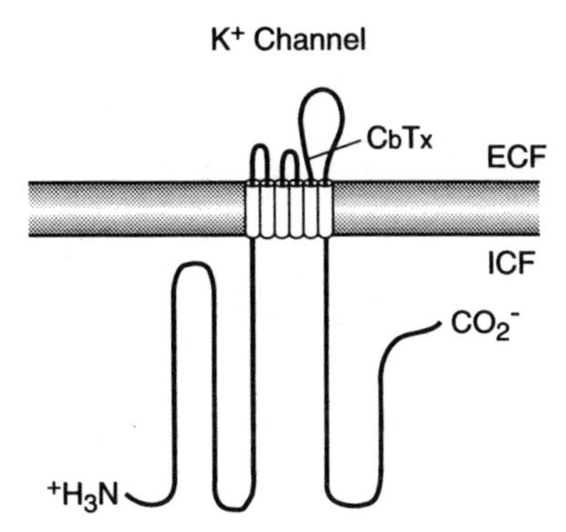

FIGURE 2–7. Postulated transmembrane looping of voltage gated K^+ channels. Charybdotoxin (CbTx) binding domain is indicated on a cytoplasmic loop. After Catterall (1988).

time constant for decay of 600 ms at 0 mV and 19° C, and with the midpoint of the K^+ channel inactivation curve being near −40 mV. In frog myelinated nerve, the K^+ channel inactivates, but nonexponentially, more slowly, and less completely than in muscle. These kinetic and regulatory differences suggest that frog nerve and skeletal muscle have different delayed rectifier channels, which could be explained by K^+ channels being coded by different genes in these two tissues. However, the microheterogeneity extends still further, and closer analysis (Hille, 1984) shows that there may be several forms of K^+ channel in the same membrane with different, but not yet fully understood, functional roles. In the first 4 years of cloning, 16 different genes for K^+ channels have been identified (Hille, 1992), and differences in structural design compared to other channels are well appreciated (Figure 2–7). Since the spatial distribution of these dominant delayed rectifier K^+ channels often is highly specific (for example, in nodes of myelinated nerves and near end plates of skeletal muscle), it is probable that isoforms could supply one means of establishing tissue-specific, within-membrane anchoring patterns.

EXCITATION-CONTRACTION (E-C) COUPLING

With the spread of the action potential across the muscle cell membrane, requiring perhaps 1 ms or less, the stage is finally set for a series of intracellular events requiring several ms to run their course and culminating in muscle contraction per se. In broad outline, these events are understood to revolve around the sarcoplasmic reticulum (SR) and Ca^{++}: release of Ca^{++} from the SR triggers contraction (see below), and active Ca^{++} uptake by the SR enables relaxation. Ultrastructural studies show that the SR of skeletal muscle is a continuous specialized membranous network composed of longitudinal tubules (LSR) that surround myofibrils and of terminal cisternae (TC) that interact via "foot proteins" with transverse tubules (TT); the latter are extremely rich in Ca^{++} channels and are responsible for storing and releasing an initial Ca^{++} pulse that correlates with and may cause further Ca^{++} release from terminal cisternae and thus, ultimately, contraction. At this point in our analysis, we wish only to examine the excitation-contraction aspect of this system (the Ca^{++} ATPase-based reaccumulation of Ca^{++} in the SR will be discussed later.), and at the outset, we must admit that this problem is not yet fully solved. However, great strides have been made in very recent studies (Figures 2–8 and 2–9). To put these in perspective, the reader should recall that several mechanisms have been advanced that could account

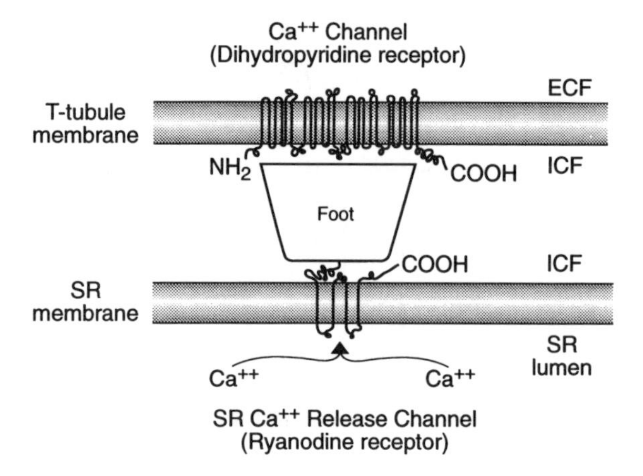

FIGURE 2–8. Current model of direct coupling of transverse tubules (TT) and sarcoplasmic reticulum (SR) through voltage sensitive Ca^{++} channels (dihydropyridine receptors) and the SR Ca^{++} release channels (ryanodine receptors). These two membrane-based protein systems are now considered to compromise key components in excitation-contraction coupling of high frequency skeletal muscles. Modified after Hille (1992).

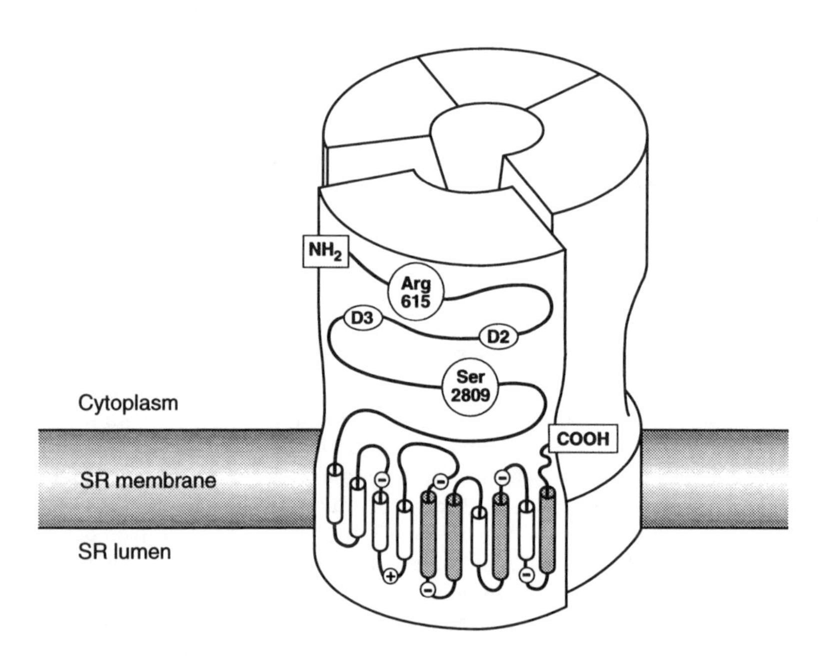

FIGURE 2–9. A current model of the SR Ca^{++} release channel (also termed the ryanodine receptor). A large N-terminal domain (about 4000 amino acids) forms the so-called ''foot'' structure in the cytoplasm. The last fifth of the molecule forms the transmembrane segments, which anchor the channel to the SR membrane, but the C-terminal loops back into the cytoplasm. A single substitution at arginine 615 to cysteine is thought to be the basis for some forms of malignant hyperthermia. Serine 2809 is the only known phosphorylation site in the cardiac ryanodine receptor. Current sequence data are consistent either with ten transmembrane domains or with four (shaded). Modified from Sorrentino and Volpe (1993).

for how the initial incoming signal (the action potential) is coupled through Ca^{++} release to muscle contraction.

The first of these is perhaps the simplest. According to this model, Na^{++} influx during the action potential (TT membrane depolarization) leads to a transient opening of Ca^{++} channels (in the transverse tubules), thus setting up the triggering cascade. Other models assume pivotal roles for H^+ concentration changes, for voltage changes across the SR membrane per se, or for surface charge effects (see Meissner, 1984). Unfortunately, none of these are comprehensive enough to be supported by and consistent with all available data. In our assessment, the model of excitation-contraction coupling currently best supported by the data *involves nucleotide-dependent allosteric opening of jSR-based Ca^{++} channels,* which may also be influenced by inositol triphosphate (IP_3). Current views of this field (Ebashi, 1991; Catterall, 1991) suggest allosteric protein–protein interactions between the TT Ca^{++} channels and the "foot proteins" of the junctional SR. At rest, when the cell membrane is polarized, the TT and the jSR do not closely interact. On depolarization, a closer interaction is initiated, presumably by the movement of a component in the TT membrane behaving like a dipole; the most likely candidate for this dipole function is the dihydropyridine sensitive TT Ca^{++} channel. The densities of these Ca^{++} channels are unusually high, even if only a small percentage of them are thought to be active as Ca^{++} channels during E-C coupling. The rest are thought to play roles as voltage sensors which, when they interact with the SR, allosterically activate Ca^{++} release channels. In skeletal muscle, this depolarization-linked dipole reorientation is thought to be the proximal cause of SR release channel activation and to lead to Ca^{++} release from the jSR for muscle activation.

Until recently, only the general characteristics of this high conductance Ca^{++} channel were known. However, with the discovery of a specific inhibitor—ryanodine (now available in isotopically labeled form)—and with current advances in the functional analysis of purified proteins reconstituted in membrane vesicles, it is already evident that the "foot proteins" are about 400,000 MW monomers that self-assemble into the "foot structures" or jSR Ca^{++} channels as tetrameres. These Ca^{++} channels display a high conductance *in vitro* as *in vivo,* compared to other types of Ca^{++} channels. Single channel (patch clamp) analyses are able to confirm earlier reports of the nucleotide dependence of SR Ca^{++} channel function: ATP binding (not ATP-dependent phosphorylation) of the foot protein increases percent open time by another approximately seven-fold factor (see Meissner, 1984; Morii and Tonomura, 1983). At this time, at least three genes for this calcium-release channel are known, hence at least three isoforms are possible (Sorrentino and Volpe, 1993).

There are two exciting features of these data. First, they come closer than ever before to explaining excitation-contraction coupling; i.e., to explaining step-by-step how, from an action potential in the TT, a voltage-sensor reorients TT Ca^{++} channels, allowing interaction with the ryanodine receptor and release of SR Ca^{++} in large enough amounts to trigger contraction within < 1 ms (measured in frog skeletal muscle at 18.5 °C). Second, and perhaps equally important, according to this model, *a true metabolic signal (ATP concentration) is a critical controlling component in the Ca^{++} release triggering contraction* (see Ebashi, 1991). Thus, the model satisfactorily explains not only excitation-contraction coupling, but it also supplies a mechanism for coupling of energy metabolism with contraction. As this model necessarily requires different kinds of Ca^{++} channel functions in at least three sites (the muscle cell membrane, the transverse tubules, and the SR), the nature of Ca^{++} channel isoforms isolated from these three sites is of particular importance.

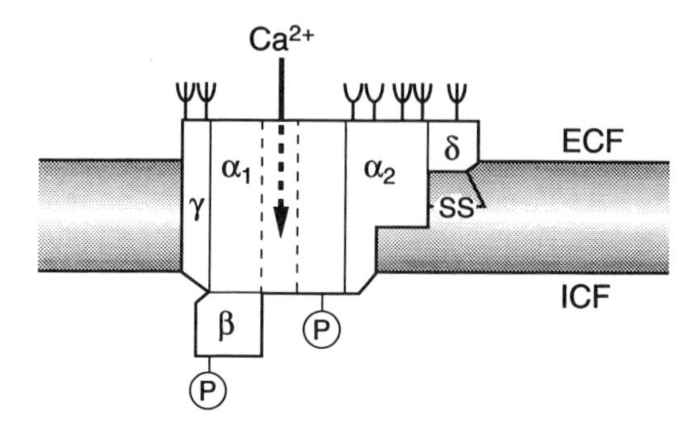

FIGURE 2–10. Current model of subunit composition of voltage regulated Ca^{++} channels. Glycosylation (ψ), phosphorylation sites (P), and disulfide bridges between subunits (-SS-) are based on biochemical data. Subunit interactions are hypothetical at this stage. Based on Catterall (1988).

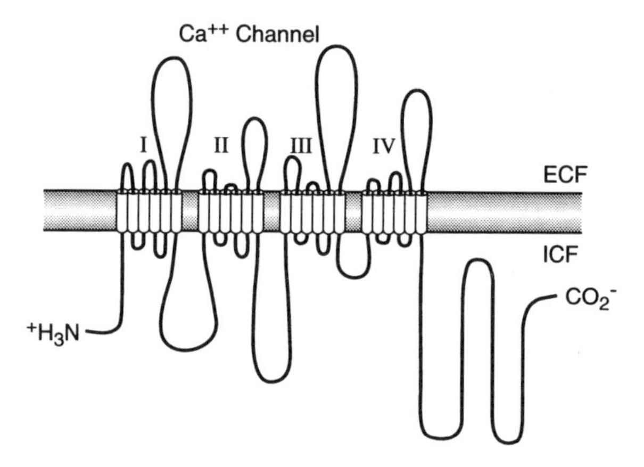

FIGURE 2–11. Currently postulated transmembrane looping of the α subunit of Ca^{++} channel. Internal repeat units are identified as I, II, III, and IV. The structure shows significant similarity to the Na^+ channel motif. Based on Catterall (1988).

Ca^{++} CHANNEL ISOFORMS

As in the examples of end-plate and Na^+ channels, great strides are now being made in correlating studies of structural (protein chemistry) and functional (single ion channel electrophysiology) properties of Ca^{++} channels (Trautwein and Hescheler, 1991; Schultz et al., 1991). One of the richest sources of Ca^{++} channels is the transverse tubules of vertebrate skeletal muscle. This Ca^{++} channel (Figures 2–10 and 2–11) is now known to consist of α-1, β, γ, and α-2-δ subunits; a family of at least six genes is known for the α and at least four for the β subunit (Catterall, 1991; Hille, 1992). On reconstitution in artificial phosphatidyl choline vesicles, they retain and express most of their *in vivo* (electrophysiologically measurable) properties. Experimental introduction into host cells shows that all subunits are required for complete expression of Ca^{++} channel function (Birnbaumer et al., 1993). Initial sequence analyses of single subunits suggest four hydrophobic (presumably transmembrane) domains homologous to similar transmembrane sequences in end-plate and Na^+ channels. The four internal repeat units also contain the positively charged S4 regions, *which are presumed to supply voltage sensitivity* (Hille, 1992; McCleskey et al., 1993). It is easy to appreciate why these domains are conservative (similar constraints on different proteins designed to span the lipid bilayer of membranes or to detect voltage change), but what is not clear is why all three channels should require four such spans of the membrane/channel subunit.

Current concepts assume that in each case the ion-specific pore is formed by subunit–subunit interactions, so 4 spans/subunit may be required for correct subunit–subunit contacts (and thus indirectly for pore formation). Analysis of TT Ca^{++} channel α-1 subunits with antipeptide antibodies reveals two different size forms: a minor 212 kDa form containing the full sequence encoded by the mRNA and a major 175 kDa form missing up to 300 amino acid residues from the carboxyl terminus. Since only a single mRNA encoding $\alpha\text{-}1_{212}$ is known, the two size forms of the α-1 subunit may arise from posttranslational proteolytic processing. Because $< 5\%$ of skeletal muscle Ca^{++} channels are active in ion conductance and $< 5\%$ of purified calcium channels have full-length α-1 subunits, it has been proposed that $\alpha\text{-}1_{212}$ may be specialized for calcium conductance, while $\alpha\text{-}1_{175}$ may serve as the voltage sensor for excitation-contraction coupling. If correct, this interpretation would mean that two protein forms encoded by the same α-1 gene may be specialized for these two separate physiological functions (see Catterall, 1991).

A point of emphasis is that the release of Ca^{++} in twitch muscles differs from other cases (such as cardiac muscle) in that Ca^{++} derives from an internal compartment (the SR) rather than from the ECF. That is, in skeletal muscle, excitation-contraction coupling relies solely on the coordinated activation of the TT and SR Ca^{++} channels; surface Ca^{++} channels open too slowly to play any significant role in the process. In contrast, in cardiac muscles, a significant portion of the Ca^{++} for excitation-contraction coupling depends on function of a cell membrane Ca^{++} channel isoform. In cardiac muscles, this surface Ca^{++} channel isoform opens faster than its homologue in skeletal muscle, and the membrane action potential is slower (Ebashi, 1991; Catterall, 1991); these are the same two reasons why this channel isoform is critical to excitation-contraction coupling in crustacean muscles as well (Hille, 1984).

In vertebrate skeletal muscles alone, then, there appear at least three groups of Ca^{++} channel isoforms: the sarcolemmal Ca^{++} channel, the TT Ca^{++} channel, and the ryanodine-sensitive SR Ca^{++} release channel. At least the first of these in skeletal muscle differs from its homologue in the heart and in nervous tissue, where (as in the presynaptic membrane) a different isoform of the Ca^{++} channel is found to occur. All three channels presumably function in excitation or contraction in cardiac muscle, but in skeletal muscle, only the TT and SR Ca^{++} channels seem to play a role in the process. Excluding the ryanodine receptors, these channels were traditionally described as L-, T-, N-, and P-isoforms (for example, see Llinas et al., 1992); however, it is now evident that the numbers of true Ca^{++} channel isoforms and their properties are so diverse that a new terminology will probably have to be introduced (Hille, 1992; Swandulla et al., 1991).

E-C COUPLING IN FAST- AND SLOW-TWITCH MUSCLES

The rate of rise and fall of twitch tension differs by a factor of 2 to 4 between fast- and slow-twitch fibers, reflecting similar differences in the rate of muscle activation. Following a single stimulation, both the action potential and the rise in the intracellular free Ca^{++} concentration are about three times faster in type-II (fast) than

TABLE 2–1
Comparison of Mammalian Fast- and Slow-Twitch Muscle

Guinea pig	Slow red (soleus)	Fast white (vastus)
T-system (% fibre volume)	0.14	0.27
jSR (% volume)	0.96	1.62
SR total (% volume)	3.15	4.99
Foot protein isoform		
Calsequestrin isoform		
Maximal rate of calcium uptake (μmol Ca mg^{-1}SR s^{-1})	5.8	22.5
Maximal calcium uptake (μmol Ca mg^{-1}SR protein)	2.8	5.5
Halftime of relaxation t/2 (ms)	113	21
Time to peak contraction (ms)	82	22
Actomyosin ATPase (μmol P_i $min^{-1}mg^{-1}$ protein)	0.05	0.13

Modified from Rüegg (1986).

in type-I (slow) fibers, suggesting that in fast-twitch muscles, calcium may be delivered more rapidly to the contractile proteins. These functional differences are matched by an adaptation of the structure of the sarcoplasmic reticulum and of the T-system. Fast-twitch muscles contain two transverse-tubular networks per sarcomere, forming contacts with the terminal cisternae of the longitudinal tubules, which release calcium into the myoplasm. Although the arrangement and architecture of the T-system is qualitatively similar in slow-twitch fibers, important quantitative differences exist (Table 2–1). Thus, the volume of the T-tubules and of the terminal cisternae and the surface area of these membrane systems all differ roughly by a factor of 2 between fast and slow fibers. Interestingly, when a fast-twitch muscle is denervated, the excitation-contraction coupling patterns change from fast to slow type, concomitantly with changing mechanical twitch characteristics, a process requiring about 3 weeks to run its course. At the same time, the composition and kind of contractile proteins remain unaltered; several weeks are required for new myosin isozymes to emerge. This suggests that it is the characteristics of the excitation-contraction coupling and rate of activation rather than the properties of the contractile machinery per se that principally determine the time course of the isometric twitch (Rüegg, 1986).

FACILITATING ROLE OF CALSEQUESTRIN

In addition to the above, fast and slow muscle activation may be influenced by the amount and kind of calsequestrin found in the jSR. Calsequestrin, a protein discovered by McLennan and Wong in 1971, is a sink for Ca^{++} in the terminal cisternae. It can bind up to 40 mmol Ca^{++} mol calsequestrin and appears to be positioned just below the "foot proteins". Thus, it is in a good position for transferring its Ca^{++} (relatively) directly to the SR Ca^{++} release channel whenever the latter is activated. The degree to which the calsequestrin–foot protein interaction speeds up Ca^{++} flux is not known. However, it is known that calsequestrin occurs in two isoforms, as two distinct gene products. One calsequestrin gene is active in cardiac (and slow-twitch) muscle, while the other is active in fast-twitch muscle. Interestingly, as mentioned above, the same is true also for the foot proteins, which also occur as tissue-specific isoforms.

3

Design of Nerve-to-Muscle Information Systems

INTRODUCTION

In Chapter 2, we noted that the transduction of nerve impulses into muscle contraction depends upon an information flow from motor nerve terminals, across an extracellular gap (the synapse) to the end plate, across the muscle cell surface, down the transverse tubules, and finally across an intracellular gap (between TT and SR) to the SR, which releases enough Ca^{++} to initiate contraction. Thus, the signal must cross (minimally) five membrane barriers and two potentially diffusion-limited gaps (one, the synapse, being extracellular, the other the intracellular gap between the TT, the SR, and the contractile elements). For these processes, at least six ion-selective transmembrane channels are required in vertebrate skeletal muscles (seven in cardiac muscle, since here a surface Ca^{++} channel isoform plays a significant role in excitation-contraction coupling). If this system were simply maximized for speed, direct electrical synapses would obviously be the signal transduction mechanism of choice, as indeed is found in fast synapses in some invertebrates. Vertebrates did not take this route, we assume because of a trade-off of regulation for speed; i.e., what seems to be selected for is a signal transmission system that is optimized for *regulated* high speed function. We have already discussed some of the regulating mechanisms known. What about speed? Since the fundamental signal transmission system depends so critically on channels, the question can be rephrased in a more general way: How fast can channels work?

CHANNELS ARE EXTREMELY EFFICIENT CATALYSTS

In considering how fast ion-specific channels work, it is worth emphasizing that two mechanisms for speeding up ion fluxes could be utilized: either the "catalytic" functions of a given channel isoform could be improved so that its turnover number (number of ions fluxed per channel per second) could be elevated, or more simply, the channel density could be increased (with no change in turnover number).

With regard to the first possibility, as catalysts designed to speed up the transit of ions (Figure 3–1), all ion channels thus far studied pass at least 1 pA of current, which corresponds to *a turnover number of* 6×10^6 *monovalent ions per second.* Most can pass much more. We do not know what the record is, but currents of between 17 and 27 pA are available for Ca^{++}-activated mammalian K^+ channels, mammalian Cl^- channels, and locust glutamate-activated channels (Methfessel and Boheim, 1982; Blatz and Magleby, 1983; Patlak et al., 1979). The maximum fluxes and conductances theoretically possible for very short pores (two atomic diameters) can be estimated using Ohm's law and Fick's law on an atomic scale and assuming

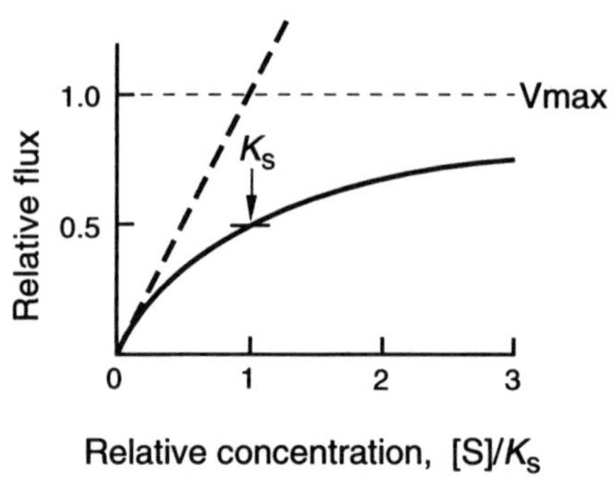

FIGURE 3–1. Michaelis-Menten kinetics for flux of ions through ion-specific channels. As the concentrations of ions on both sides of the pore are increased, ionic flux increases asymptotically to a maximum value. [S] refers to ion concentration; K_s is the concentration for half maximal flux; Vmax is maximum flux.

that ionic mobilities in the pore are the same as in free solution. In round numbers, the limits in 120 mM salt solution are a maximum of about 33 pA of current at 0 mV, and a conductance of 300 pS. This theoretical maximum flux can be expressed more generally as a second-order rate constant of 1.9×10^9 ions M^{-1} s^{-1}, close to diffusional limitations (Hille, 1984).

There are a number of reasons why most channels do not reach these theoretically maximum turnover rates. For example, ions change their state of hydration in channels, and the time course of ligand substitutions (dehydration with groups on the pore acting as surrogate water molecules) can act to slow down permeation. Indeed, the removal of water at a narrow region of the channel is considered a rate-limiting step and one that gives some ion channels some of their ionic specificity (Hille, 1984; 1992).

A second reason why channel fluxes may be less than theoretically expected is because most channels show saturation kinetics. In the Michaelis-Menten model of enzyme kinetics, the velocity of reaction reaches a saturating value (V_{max}) at high substrate concentration because substrate molecules compete for binding to active sites, and each enzyme takes a finite time to convert the bound substrate into products and to release them. Similarly, ionic channels can be regarded as catalysts with a limited number of binding sites and taking a finite time to process their substrates.

In addition to retardation due to dehydration and saturation, ion fluxes can be slowed down because of ionic flux coupling (due to more than one ion in the channel at once), electrostatic barriers, or even physical barriers. Any of these factors could arise simply from the fact that most "real" channels are substantially larger than the short (two-atomic diameter) pore length assumed for calculating the theoretical maximum flux rates. Nevertheless, some channels (for example, SR Ca^{++} channels) can come within about 1.5 orders of magnitude of theoretical maximum fluxes calculated for short pores (see Hille, 1984; 1992). In terms of ion flux rates, such channels therefore are close to or at a state of adaptational perfection: further speeding up the catalytic component of such channels would not be reflected in increased flux rates.

This impression is also reflected in comparisons of these flux capacities with other processes. Such analysis shows that in metabolic or physiological terms, channels can support relatively awesome flux rates: *very few enzymes and no known ion exchangers, ion carriers, or ion pumps even approach such speed.* In fact, Hille (1984; 1992) argues that estimates of speed alone can serve as major evidence that biological ionic channels are pores. This can be illustrated by comparisons of turnover numbers.

TABLE 3–1
Turnover Numbers of Representative Enzymes, Channels, Transporters, and Pumps

	Substrate	Substrate turnover (no./s)	Temp. (° C)
Enzymes			
Catalase	H_2O_2	5×10^6	20
Carbonic anhydrase	CO_2	1.4×10^6	25
5-3-Ketosteroid isomerase	Steroid	7.3×10^4	25
Acetylcholinesterase	ACh	1.6×10^4	25
H_4LDH	Pyruvate	43×10^3	30
M_4LDH	Pyruvate	12×10^3	30
Most *enzymes*	Organic molecule	$0.002–1.0 \times 10^4$	
Most channels	ions	10^6 to 10^9	
Carriers			
Cl^-/Cl^- exchange	Cl^-	5×10^4	38
Valinomycin	Rb^+	3×10^4	23
Trinactin	NH_4^+	4×10^4	23
Na^+-K^+ ATPase	Na^+	5×10^2	37
Ca^{2+} ATPase	Ca^{2+}	2×10^2	
Glucose transporter	Glucose	$0.1–1.3 \times 10^4$	38

From Hochachka, P. W. and Somero, G. N. (1984) *Biochemical Adaptation.* Princeton University Press; Hochachka (1988a).

In enzymology, the turnover number is defined as the maximum number of moles of substrate processed per mole enzyme per second, which is taken as a measure of the maximum catalytic capacity of an enzyme. Table 3–1 lists turnover numbers for some of the faster known enzymes. By far the fastest is catalase, which converts its simple substrate, HOOH (hydrogen peroxide), to O_2 and H_2O in 200 ns. Carbonic anhydrase is next and can hydrate CO_2 in under 700 ns. The enzyme, 3-ketosteroidisomerase, takes 10 μs to catalyze the transfer of a proton between two carbon atoms. Acetylcholinesterase is also an unusually fast enzyme, requiring only 60 μs to hydrolyze acetylcholine. By far the majority of enzymes seem to have lower turnover numbers, in the range of 20 to 10^4 substrate molecules per second. The Q_{10} of enzymatic catalysis is typically near 3.0, corresponding to activation energies of 18 kcal/mol; channels by comparison display Q_{10} values of about 1.5 or less, similar to diffusion; values higher than this sometimes are observed and imply rate determining processes such as allosteric transitions or other conformational changes, which are known from other systems to display high temperature coefficients. Transporters, pumps, and carriers appear to be rather similar to metabolic enzymes, and their turnover numbers are in no case known to date even close to the capabilities of an aqueous channel (Table 3–1). What this suggests is that very little room is left for upward adjustments of the catalytic capacities of ion channels; as a means for attaining even higher flux rates, further tuning up of the catalytic mechanism per se therefore would seem to be relatively fruitless. We are thus left with the channel densities option.

CHANNEL DENSITIES ARE USUALLY RATHER LOW

To put this problem into perspective, it is important to realize that channels are not a major chemical constituent of membranes. This can be shown by considering how many protein macromolecules can fit into a membrane. This maximum may be

TABLE 3–2
Adaptability of Channel and Pump Densities

	Observed density in membrane (no./μm²)	Molecular mass (kDa)	Density for hypothetical 250-kDa protein units (no./μm²)
Vertebrate rhodopsin in rod disks	30,000–64,000		
Halobacterium rhyodopsin	50,900	3 × 26	15,900
Ca^{++} ATPase (heater organ)	34,000	2 × 100	8,500
Rat hepatocyte gap junctions (connexons)	27,700	6 × 26	17,300
Rhodopseudomonas viridis photosynthetic reaction center	15,000	535	32,000
ACh receptor, motor end plate	10,000	250	10,000
Na^+ channel (node of Ranvier, rat myelinated nerve)	23,000	316	18,200
Na^+ channel (rat skeletal muscles)	200–600	316	160–470
Na^+-K^+ ATPase, chloride cells of fish gills	8,000	2 × 135	8,600
Na^+-K^+ ATPase, striated muscles	1,600–3,350	2 × 135	1,700–3,700
Na^+-K^+ ATPase, tracheal epithelium	2,400	2 × 135	2,600
Na^+-K^+ ATPase, RBC	1	2 × 135	1
Ca^{2+} ATPase, rabbit sarcoplasmic reticulum	8,700	2 × 102	7,100

Date from Hochachka (1988a) and Block (1991).

compared with densities of the most densely crowded membrane proteins known, which range up to 64,000 per μm² (Table 3–2). To make this estimate easily comparable to densities of channels, the observed densities of proteins are converted into equivalent densities for model 250-kDa units (Table 3–2). This is done by assuming, for example, that the space taken up by 250 proteins 100 kDa in size is equivalent to a space required for 100 proteins 250 kDa in size. Such equivalent densities range up to 32,000 proteins per square micrometer. By comparison, the channel densities of such critically important proteins as muscle Na^+ channels occur at only 2 to 6 × 10^2 per μm² (Table 3–2). Densities some *100 to 10,000 times lower than theoretically possible are not at all unusual for numerous ion-specific channels and are only possible because of the very high catalytic (ion-flux) capacities of channels* (Hille, 1984; 1992). However, in some cases, high catalytic capacities on their own are not enough; that is where the channel density option is really required.

CHANNEL DENSITIES MAY BE ADJUSTED UPWARD

The best known examples of upward adjustments in channel densities are Na^+ channels along transverse tubules in skeletal muscle, Na^+ channels at the Nodes of Ranvier in myelinated nerves, and AChR channels at end plates in skeletal muscle, *all three being sites requiring highest Na^+ fluxes.* In the Nodes of Ranvier, Na^+

channels are seemingly very concentrated, yet there evidently remains a lot of room for many other membrane proteins (Table 3–2). In the case of end-plate channels, the post-synaptic end-plate junctional folds are so packed with AChR proteins that they are estimated at about 30% of the upper theoretical limit. At the limit, the membrane, of course, would be solid protein, so the theoretical maximum actually is an overestimate. What this means is that for AChR channels at any rate there again is very little room left for any further upward adjustment in channel densities (perhaps another 1.5–2-fold at maximum, and possibly less).

Thus, two instructive insights arise from our general analysis: first, channels for ion transmission are extremely efficient, high-capacity catalysts and so can often sustain normal functions at low densities. However, we learn second that where needed, channel densities can be increased to fill about one-third or more of the maximum membrane volume available to them. Now let us see how these two adjustment strategies are used at each step in the overall nerve-to-muscle signaling system.

DESIGN CRITERIA FOR PRESYNAPTIC SIGNALING PROCESSES

In order for the presynaptic Ca^{++} channel isoform to serve a useful role in transducing the signal for release of ACh, it must fulfill a set of minimum requirements. First, it clearly should be and is capable of sensing arriving motor neuron impulses, which are the most immediate indicators of the need for action; voltage regulation of Ca^{++} channel opening is in fact common to all Ca^{++} channel isoforms.

Second, because Ca^{++} can have devastating effects at high cytosolic concentrations (see Hochachka, 1986), the Ca^{++} channel-mediated signal should be amplified. Knowing the time course of presynaptic membrane activation (about 200 ms), the turnover number of Ca^{++} channels (at least 10^6 ions per channel per second), the number of ACh-containing vesicles that are released per impulse (100 to 300), and the number of acetylcholine molecules per vesicle ($\sim 10^4$), it is estimated that the Ca^{++} signal is amplified by about two orders of magnitude. If we assume that the fewer the number of Ca^{++} ions required for ACh release the better, then amplification would be achievable in one of three ways: (i) the Ca^{++} affinity of the Ca^{++} target (thought to be a fusion-inducing protein) could be increased, triggering exocytosis at lower Ca^{++} pulses; (ii) the number of ACh vesicles/impulse could be elevated; and (iii) the amount of ACh/vesicle could be increased. In both the latter two mechanisms, a smaller amount of Ca^{++} would be required for a given ACh signal; i.e., amplification would increase.

A third, and most important, criterion that must be met by presynaptic signals is speed; i.e., the speed of the Ca^{++} channel response to arriving impulse should be at least fast enough so as not to limit the overall flow rate of information. Because we know that Ca^{++} channels work very rapidly, we would not anticipate any adjustments in turnover rates for channels in fast muscles. Instead, we would expect the densities of these Ca^{++} channel isoforms in fast-twitch muscles to exceed the densities in presynaptic membranes of slow muscles, and upward density adjustments should be greatest in superfast muscles (such as eye muscles, jaw muscles, fish sonic muscles, and insect sonic muscles).

In contrast, downward density adjustments should lead to slower, weaker, or more fatigable muscles. Such consequences of downward adjustments are now well described at least in one case: the Lambert-Eaton syndrome. Abbreviated LES, this syndrome is a rare myasthenic condition characterized by muscle weakness and excessive fatigability. Recent research indicates that LES involves an antibody autoimmune response, in which the *presynaptic Ca^{++} channel is the target antigen. The functional or effective Ca^{++} channel densities of presynaptic fibers are compromised as a result, and ACh release at the neuromuscular junction is seriously impaired* (Lindstrom, 1985).

In our context, the lesson to be learned is that upward or downward adjustments in the density of isoforms of Ca^{++} channels in the presynaptic membrane are evidently required in the adaptation of muscles for specialized capabilities.

DESIGN CRITERIA FOR POSTSYNAPTIC SIGNAL TRANSDUCTION

In design terms, there are two critical functions required at the post-synaptic membrane: reception of signal and transmission of signal. As we have seen, these two jobs are elegantly incorporated into one structure, for the AChR protein is at once an ACh receptor and a channel, with channel opening being dependent on binding two ACh molecules. In principle, the efficiency of these functions could be improved by upward adjustments in either (i) the catalytic capacity of the channel or (ii) the channel density at junctional folds. As the former is already considered to be pretty well maximized, we do not consider that this feature remains an effective site of selective pressure. However, at least in part because of the impressive speed of AChR channels, relatively small changes in density may indeed lead to large changes in effective rate at which end plates can be depolarized. There are two aspects to how density adjustments bring this about: (i) sensitivity of the response to transmitter (ACh) and (ii) conductance change per se. Katz and Miledi (1972) and Anderson and Stevens (1973) showed that each activated ion channel contributes a fixed conductance change and therefore more current flows in direct proportion to the number of ion channels activated. Since the sensitivity in signal (ACh) is a complex competitive interplay between ACh binding by AChR and AChE (which inactivates the signal), the relationship between AChR end-plate densities and ACh sensitivity might in principle be relatively complex. Interestingly, this does not turn out to be the case. Land et al. (1977) showed that *muscle end-plate sensitivity to ACh-release is in fact directly proportional to change in AChR end-plate densities*. However, to our knowledge, the question of how AChR and AChE densities coadapt remains unanswered.

A particularly dramatic example of the impact of *reduced* AChR channel densities at muscle end plates comes from studies of the muscular weakness and fatigability characterizing a disease known as myasthenia gravis (MG). Pathological changes at neuromuscular junctions caused by MG include antibodies and complement bound to the postsynaptic membrane and a loss in AChR density; as a result, the structures of junctional folds of postsynaptic membranes are also simplified. As in the Lambert-Eaton syndrome discussed above, MG is thought to be caused by an antibody-mediated autoimmune response to acetylcholine receptors; i.e., antibodies self-directed at an internal target: AChR channels in muscle end plates. Current understanding of this unfortunate disease is reviewed by Lindstrom (1985). In our context, these recent studies of MG emphasize the validity of our conclusion that an important mechanism for speeding up signal transmission at the AChR step in the information flow from nerve $\rightarrow$ muscle is upward *adjustment in AChR channel density* at muscle end plates. That is why in fast-twitch vertebrate muscles, AChR end-plate densities exceed those in slow muscles and why we would expect even higher densities in superfast muscles, such as the ciliary muscles of the eye or jaw muscles of predators. In fact, in general, we would expect that the faster the muscle, the higher the end-plate AChR densities:

insect sonic muscles; 500 Hz	>	fish sonic muscles; about 380 Hz	>	synchronous insect flight muscles; about 100 Hz	>	vertebrate fast muscles; about 40 Hz

If this relationship is direct, as suggested by Land et al. (1977), then we would expect about a ten-fold difference in AChR end-plate densities between vertebrate fast muscles and the fastest muscles currently known, insect sonic muscles (Josephson and Young, 1985). Unfortunately, there is the rub. We have already argued that even in the relatively sluggish muscles of the frog, end-plate densities come to within about 30% of the theoretical maximum. Given that ACh sensitivity and end-plate conductance changes depend directly on AChR channel density, it follows that by density changes alone, nature could tune up transmission rates at this step *maximally by three-fold, not the ten-fold observed.* Therefore, something is missing in our analysis to this point. We suspect that the missing element is probably acetylcholinesterase (AChE) ''density'', which must somehow influence the sensitivity and timing of signal transmission. From current theory, we assume that for a given AChR density, *decreasing AChE titers should allow a greater conductance change per given number of ACh molecules* arriving at the postsynaptic membrane. However, we know of no work that has tried to quantify the relationship between AChR densities, AChE enzymic titers, ACh sensitivity of the end plate, and conductance changes of the end plate to a given acetylcholine signal. This is an area that is clearly in need of more research.

DESIGN CRITERIA FOR Na^+ CHANNEL FUNCTIONS

As in the examples of the presynaptic isoform of Ca^{++} channels and the muscle forms of AChR channels, there are several criteria which must be met for effective Na^+ channel function during muscle excitation. In the first place, the Na^+ channel (or any channel playing this role, e.g., Ca^{++} channel in crustacean muscles; Hille, 1992) should be responsive to voltage change. As we have seen, all Na^+ channel isoforms are in fact voltage regulated, with an intrinsic voltage sensor that is a part of the channel protein structure (probably a strongly positively charged polypeptide spanning the sarcolemma). So far, Na^+ channel voltage sensors seem to be 100% conserved, wherever they have been studied; thus, we conclude that this is not an adaptable Na^+ channel function.

Another key design criterion is speed. The Na^+ channel appears to be within about 2 orders of magnitude of the maximum catalytic potential possible for very short channels; this theoretical ceiling must necessarily be lower for longer (real life) channels; thus, Na^+ channels are judged to be close to kinetic perfection. That is why further upward adjustment along these lines is considered unlikely.

The only other way to increase Na^+ flux rates when the Na^+ channel is in open configuration is to increase channel density. This turns out to be the main adaptable component of Na^+ channel function. As indicated in Table 3–2, Na^+ channel densities are highest where highest Na^+ fluxes are required. Thus, it is expected and indeed observed that in muscle, highest Na^+ channel densities are to be found localized near end plates and near transverse tubules; that is also why superfast muscles (ciliary muscles of the eye have been studied) display higher Na^+ channel densities than fast skeletal muscles, which in turn have higher densities than in slow muscles. The hierarchy should be the same as for AChR channels: insect sonic muscles > fish sonic muscle → insect synchronous flight muscles > vertebrate fast muscles > vertebrate slow muscles. What we do not know at this time is whether or not Na^+ channel densities are adjustable on shorter term basis by training, although it is possible that fiber-type transitions with training would automatically achieve this goal.

DESIGN CRITERIA FOR TT AND SR Ca^{++} CHANNELS

The pattern of our analysis should now be gradually becoming evident: The faster the muscle under consideration, the faster must be the information flow capacity from nerve to muscle, and nowhere is this more evident than in the adaptations of SR. For the same reasons as already argued in the sections above, the main mechanism utilized for tuning up Ca^{++} fluxes through the TT and SR Ca^{++} release channels appears to be adjusting SR volume density and channel densities per unit of membrane surface in each case. Thus, it is well known that TT membranes are one of the richest sources of Ca^{++} channels and that fast muscles have more TT and SR membrane surface than slow muscles (Table 2–1). To our knowledge, no comparison has been made of the densities of the SR Ca^{++} channel isoforms, but it is a fair guess that their cellular content rises in proportion to the increase in SR surface area in fast muscles.

In considering fast muscles, it is important to emphasize that most vertebrate fast muscles are either largely glycolytic or glycolytic-oxidative (see Chapter 6). However, some less well-studied superfast vertebrate muscles, such as the eye muscles, are notable not for their anaerobic capacities but for their aerobic ones. In the fastest muscles known, sonic muscles of some insects (Josephson and Young, 1985), oxidative metabolic capacities are greatly elevated. Whether or not the source of ATP for the contractile machinery is an anaerobic or aerobic pathway, however, all fast muscles seem necessarily to have greatly expanded SR and thus expanded densities of both TT and SR channels. In the case of sonic muscles of insects (Josephson and Young, 1985), SR volume density may approach one-third of each muscle cell. Since mitochondrial abundance is also elevated, in these kinds of muscles, each sarcomere is only about one-third myofilaments; the rest of cell volume is taken up with SR and mitochondria, implying an interesting trade-off, *speed for strength.*

Finally, perhaps the most fundamental adaptation of skeletal muscle for facilitating the speed of muscle function is in the inherent design of the excitation-contraction coupling mechanism. The best way to explain this is through comparison of the EC coupling in cardiac vs. skeletal muscle. In the former, the main source of Ca^{++} for EC coupling is the ECF, and the main route of entry is the sarcolemmal-based Ca^{++} channel. In skeletal muscle, this route of entry is bypassed, and the role of Ca^{++} as an intracellular second messenger for EC coupling is minimized or completely lost. Instead, EC coupling now depends primarily on direct protein–protein interactions (TT Ca^{++} channels as voltage sensors interacting with the SR Ca^{++} release channels). This mechanism of EC coupling is considered to be analogous to allosteric mechanisms in enzyme regulation and is thought to contribute to the rapid contraction characteristics of skeletal muscles relative to cardiac muscles (Catterall, 1991). Interestingly, new molecular biological studies of the dependence on the sarcolemmal Ca^{++} channel for EC coupling have localized the molecular basis for the skeletal muscle—heart muscle differences. The point of departure for these studies is the well-known requirement for ECF Ca^{++} in EC coupling in cardiac muscle. A similar requirement for ECF Ca^{++} is found when the α-1 subunit of the cardiac sarcolemmal Ca^{++} channel is expressed in skeletal muscle myocytes from muscular dysgenesis mice to restore EC coupling. Evidently, the mode of EC coupling (allosteric vs. Ca^{++} as second messenger) is determined by the α-1 subunit of the Ca^{++} channel. The construction of a series of chimeric Ca^{++} channels (composed of different doses of either heart-type or skeletal muscle-type Ca^{++} channels) indicates that a large intracellular loop in α-1 subunit connects two homologous domains (II and III) and determines the direct allosteric EC coupling mechanism of skeletal muscles. It is believed that this segment of the α-1 subunit of the TT Ca^{++} channel (directly or through an additional protein) interacts with the Ca^{++} release channel (Catterall, 1991). These exciting developments not only help us to understand how the allosteric EC coupling mechanism works and of course why it is faster than the second-messenger system found in cardiac EC coupling, but they also supply guidelines for detailed comparisons of the same problem in different fiber types and especially in different species, where these processes are known to be even further fine-tuned for even faster function.

OVERALL DESIGN PRINCIPLES FOR NERVE-TO-MUSCLE INFORMATION FLOW SYSTEMS

From the discussion in this chapter, it is possible to discern several functional constraints or rules for design of nerve-to-muscle information flow systems. First, the catalysts involved are to a large degree ion-specific channels, whose function is near kinetic perfection; thus any upward regulation of catalytic function does not involve further increase in catalytic efficiencies or turnover numbers of channels involved. Second, although regulation of ion channels is not yet well understood, the information available does imply that regulation mechanisms and catalytic functions of channels are fairly conservative (i.e., similar in widely differing species and tissues). Third, because structure, catalysis, and regulation are all highly conserved traits of channels, tuning up speed of function often involves simply changes in the amounts (or functional densities) of channels, with fast muscles typically displaying higher densities than found in slow muscles. Sometimes, as in the case of the jSR "foot proteins" (Ca^{++} release channels), adjustment in Ca^{++} flux capacity involves amplification of SR membrane. However, because of the widely differing catalytic efficiencies of jSR Ca^{++} channels and the SR Ca^{++} ATPases (whose Ca^{++} fluxes must be approximately balanced), there is little doubt that the SR structural amplification must decidely favor the Ca^{++} ATPase regions.

A fourth design principle also arises from the high inherent catalytic efficiency of channels, namely, that they usually do not need to be packed in at high densities. However, there are some exceptions to this general rule, where specific high-speed functions (high specific flux rates of ions) are so important that channel densities are pushed toward upper limits (e.g., Na^{+} channels at Nodes of Ranvier and at muscle end plates).

Finally, in a few cases, it is evident that the function of information transfer depends upon coadaptation of more than one component at once. AChR (end-plate channel) densities, for example, are necessarily coadapted with AChE catalytic activities, although the exact nature of selective forces here is not understood. Similarly, the jSR foot proteins or Ca^{++} release channels clearly must coadapt first with TT Ca^{++} channels in order to form the allosteric EC coupling mechanism and then second with Ca^{++} ATPase catalytic capacities so that Ca^{++} fluxes during muscle activation and contraction can be exactly balanced by Ca^{++} fluxes in the opposite direction during muscle relaxation. Again, much research is still required to sort out selective forces that are in action in such cases.

4

Energy Demand of Muscle Machines

INTRODUCTION

In considering events following arrival of the acetylcholine signal at the muscle end plate, it is convenient to look at muscle work as arising from the operation of the classical adenylate cycle of bioenergetics:

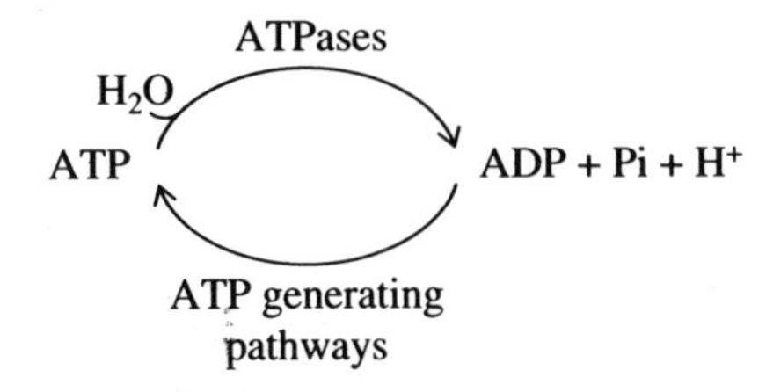

While the cycle operates in all working muscles, the rate of operation (the rate of ATP turnover) and hence the rate of work output depends upon the kind of muscle involved, because muscles are tuned through adaptation for specific kinds of performance. Some muscles, for example, are designed for speed; one of the fastest muscles currently known is the sonic muscle of certain insects, where *sustained* function at over 500 Hz is achievable (Josephson and Young, 1985). Its ultrastructure (which is well suited for very high rates of ATP turnover) could hardly be more different from that of vertebrate muscles designed for short-term high power output. In vertebrates, the latter are termed fast-twitch glycolytic (FG or Type IIb) muscles to contrast them with fast-twitch glycolytic oxidative (FOG or Type IIa) and with slow-twitch oxidative (Type I) muscles. Hummingbird flight muscles formally are FOG or Type IIa muscles, yet are also capable of rather fast, sustained function; they show ultrastructural features that are intermediate between the above two extremes (Suarez et al., 1991).

We do not know the ATP turnover rate that is sustainable by insect sonic muscles, but estimates for hummingbird flight muscles surpass 600 μmol ATP g muscle^{-1} min^{-1} or a metabolic rate that is some 20 times higher than for human leg muscles during long distance running. In contrast to sustainable work, short-term (anaerobic) ATP turnover rates in mammalian muscles, including those of man, approach 500 μg ATP muscle^{-1} min^{-1}, so at least for short time periods, our fast muscles can nearly match those of hummingbirds on a gram-for-gram basis.

TABLE 4–1
Metabolic Rates and Actomyosin ATPase Turnover Rates during Various Kinds of Muscle Work

Muscle type	ATP turnover or metabolic rate, μmol/g wet wt/min	Actomyosin turnover rate (corrected for Ca^{++} ATPase)[a]
Insect flight	1920 (sustained)[f]	—
Hummingbird	605 (sustained)	75[b]
Mammalian fast-twitch	480 (short term phosphagen-based)	30[c]
Human knee extensors	120 (maximum sustainable)	7.5[d]
Human leg muscles	30 (sustained running, e.g., marathon)	1.8[e]

[a] Assuming 25% of ATP turnover rate is due to Ca^{++} ATPase function.
[b] Assuming myosin concentration is $5 \times 10^{-5}M$, about one-third that in mammalian fast-twitch muscle (Yates and Greaser, 1983), as implied from ultrastructural studies (see Mathieu-Costello et al. 1992b).
[c] Assuming myosin concentration is $2 \times 10^{-4}M$ (Yates and Greaser, 1983) and that 40 μM of high energy phosphate are hydrolyzed per gram muscle in 5 s.
[d] Assuming 90% of whole-organism O_2 uptake is used by these muscles at the individual's maximum work rates (Saltin, 1985).
[e] Assuming an average myosin concentration of $2 \times 10^{-4}M$ (Yates and Greaser, 1983) and metabolic rates given by McGilvery (1975).
[f] From Casey et al. (1992) for a 0.1 g bee in flight.

As we shall see below, it is possible to isolate and purify skeletal muscle contractile proteins, and for mammalian preparations, the accepted turnover numbers are about 10 μmol ATP μmol actomyosin^{-1} s^{-1}. From Table 4–1, it is evident that during very heavy but sustainable work of the knee extensor in man, this theoretical maximum value can be approached; as there is no reason to suppose that knee extension is in any way unusual, we can assume that ATP turnover rates approaching theoretical maximum values may be commonly achieved during muscle work in man. They clearly can be surpassed by a factor of perhaps up to several-fold during short-term burst work in most mammals. In smaller, faster animals, such as the hummingbird, maximum actomyosin ATP turnover rates expected from *in vitro* measurements can be surpassed by nearly an order of magnitude!

How can we rationalize these differences in structure and function of muscles? In particular, how can we explain *in vivo* catalytic rates that are higher than actual turnover numbers observed *in vitro*? To understand how this apparent paradox arises and where the relationship between structure and function in muscle is adjustable, we must begin with an examination of our current theories of muscle composition and contraction.

THE SARCOMERE: THE BASIC CONTRACTILE UNIT

The fundamental structure of muscle is well described in many biochemistry and physiology texts and reviews, and therefore in the next sections we will present only a relatively brief discussion of this topic. Although we feel that readers already familiar with this section may skip it completely, we consider that knowledge of the basic structural data is essential to our later arguments.

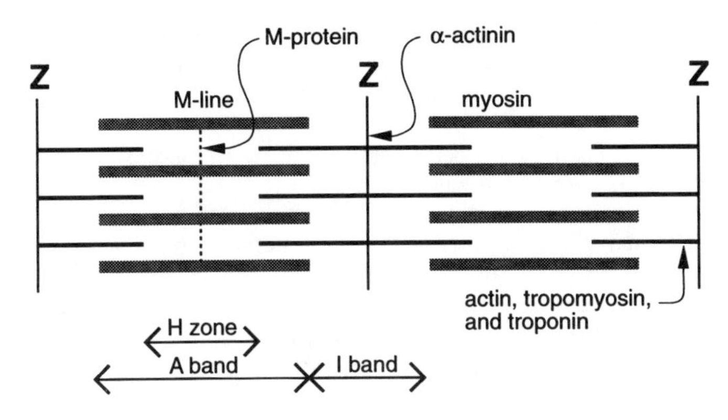

FIGURE 4–1. Schematic diagram of the sarcomere of skeletal muscles, identifying biochemical components and the various bands that can be observed in electron micrographs.

For many years, it has been known that all vertebrate skeletal muscles display a striated appearance when examined under the light microscope. Such muscles consist of cells, each of which is surrounded by an electrically excitable membrane called the sarcolemma. A muscle cell contains many parallel myofibrils, each about 1 μm in diameter.

Electron microscopy of many different kinds of skeletal muscles reveals a common functional unit, called the sarcomere, repeating every 2.3 μm (23,000 Å) along the fibril axis. A dark *A* band and a light *I* band alternate regularly. The central region of the *A* band, termed the *H* zone, is less dense than the rest of the band. A dark *M* line is found in the middle of the *H* zone. The *I* band is bisected by a very dense narrow *Z* line, a kind of proteinaceous coupling of adjacent sarcomeres (Figure 4–1).

Electron micrographs of cross-sections of myofibrils show that there are two kinds of interacting protein filaments. The thick filaments have diameters of about 150 Å, whereas the thin filaments have diameters of about 70 Å. The thick filaments primarily consist of myosin. Thin filaments contain actin, tropomyosin, and troponin. Alpha-actinin is present in the *Z* line, whereas an *M*-protein is located in the *M* line.

The *I* band consists of only thin filaments, whereas only thick filaments are found in the *H* zone of the *A* band. Both types of filaments are present in the other parts of the *A* band. A regular hexagonal array is evident in cross-sections, which show that each thin filament has three neighboring thick filaments, and each thick filament is encircled by six thin filaments. The thick and thin filaments interact by cross-bridges, which, as we shall see, are domains of myosin molecules. Cross-bridges emerge at regular intervals from the thick filaments and bridge a gap of 130 Å between the surfaces of thick and thin filaments.

SLIDING FILAMENT MODEL OF CONTRACTION

Muscle shortens by as much as a third of its original length as it contracts. How can we explain this shortening? Some three decades ago, two laboratories (the first headed by A. Huxley and R. Niedergerke, the second by H. Huxley and J. Hanson) independently proposed a sliding filament model of contraction based on X-ray, light microscopic, and electron microscopic studies. There are three essential features in this model:

1. Thick and thin filament lengths do not change during muscle contraction.
2. Sarcomere length decreases during contraction because the two types of filaments overlap more; i.e., thick and thin filaments are moved past each other in contraction.
3. Force is generated by a process that is coupled to the movement of thick and thin filaments past each other.

The most fundamental support for this model comes from measurements of the lengths of the *A* and *I* bands and of the *H* zone in stretched, resting, and contracted muscle. First, the length of the *A* band is constant, which means that the thick filaments do not change size. Second, the distance between the *Z* line and the adjacent edge of the *H* zone is also constant, which indicates that the thin filaments do not change size. Third, the *H* zone and the *I* band decrease in size on contraction, because the thick and thin filaments overlap more.

THREE FUNCTIONS OF MYOSIN

Myosin has three important biological activities. First, myosin molecules spontaneously assemble into filaments in solutions at physiological ionic strength and pH; thick filaments in fact are formed mainly by myosin molecules. Second, as already emphasized above and first discovered by Engelhardt and Lyubimova, myosin is an ATPase capable of catalyzing the reaction:

$$ATP + H_2O \rightarrow ADP + P_i + H^+$$

This reaction is now considered the immediate source of the free energy that drives muscle contraction. Third, myosin binds to the polymerized form of actin, the major constituent of the thin filament. This interaction is the main basis for the generation of the force that moves the thick and thin filaments relative to each other.

Myosin is a very large molecule (500 kDa). It contains two identical major chains (200 kDa each) and four light chains (about 20 kDa each). Electron micrographs show that myosin consists of a double-headed globular region joined to a very long rod. The rod is a two-stranded α-helical coiled coil.

TAKING MYOSIN APART TO IDENTIFY FUNCTIONAL DOMAINS

Much of our understanding of the biochemical properties of myosin depends upon the fact that different fragments obtained following controlled proteolysis retain different partial activities of the intact molecule. This approach was first initiated by Szent-Gyorgyi, who showed in 1953 that myosin is split by trypsin into two fragments, called light meromyosin (LMM) and heavy meromyosin (HMM). Light meromyosin, like myosin, forms filaments. However, it lacks ATPase activity and does not combine with actin. Electron micrographs indicate that LMM is a rod, which is why LMM solutions display high viscosity. Optical rotatory properties of LMM are consistent with an α-helix content of 90%. Furthermore, X-ray diffraction patterns of LMM fibers display a strong reflection at 5.1 Å, which is characteristic of an α-helical coil. For these kinds of reasons, it is widely held that LMM is a two-stranded α-helical rod for its entire length of 850 Å. Heavy meromyosin (HMM) has quite different properties: for example, it catalyzes the hydrolysis of ATP and binds to actin, but it does not form filaments. HMM consists of a rod attached to a double-headed globular region. It can be split by further proteolysis into two globular subfragments (called S1) and one rod-shaped subfragment (called S2). *Each S1 fragment contains an ATPase active site and a binding site for actin.* Furthermore, the light chains of myosin are bound to the S1 fragments.

GLOBULAR AND FILAMENTOUS FORMS OF ACTIN

The major constituent of thin filaments is actin. In solutions of low ionic strength, actin is a 42-kDa monomer called *G-actin* because of its globular shape. As the ionic strength is increased to the physiological level, G-actin self-assembles into a fibrous form called F-actin, which closely resembles thin filaments. An F-actin fiber looks in electron micrographs like two strings of beads wound around each other. X-ray diffraction patterns show that F-actin is a helix of actin monomers, with a diameter of about 70 Å. The structure repeats at intervals of 360 Å along the helix axis.

The complex formed when a solution of actin is added to a solution of myosin is termed actomyosin, and this complex formation is accompanied by a large increase in the viscosity of the solution. Over four decades ago, Szent-Gyorgi showed that this increase in viscosity is reversed by the addition of ATP; i.e., *ATP dissociates actomyosin* to actin and myosin. Szent-Gyorgyi also showed that threads of actomyosin could be formed by flow-orienting them in solution. When such actomyosins are exposed to ATP, K^+ and Mg^{++}, the *threads contract,* whereas threads formed from myosin alone do not. Such experiments unequivocally demonstrate that interactions between myosin, actin, and ATP are necessary for contraction. These interactions involve a fascinating interplay between actomyosin as an enzyme and myosin as a mechanical transducer.

ATPase COUPLING WITH FILAMENT MOVEMENT

The ATPase activity of myosin is markedly enhanced by stoichiometric amounts of F-actin, the turnover number increasing some 200 times, from 0.05 s^{-1} to 10 s^{-1}. Although the means by which this is achieved are in fact very complex, the ATPase catalytic cycle can be simplified to four dominant steps (Figure 4–2), which are consistent with (and based upon) biochemical, electron microscopic, and X-ray diffraction data. According to this model, each S1 head in resting muscle occurs in a state detached from thin filaments; in this state, myosin contains tightly bound ADP + P_i, and S1 heads form an ordered helical array around the thick filament. On stimulation, each S1 head becomes attached to the thin filament in a perpendicular orientation. ADP and P_i bound to S1 are then released, and the S1 head adopts a tilted orientation. The conformational change (high → low energy state) associated with the release of bound ADP and P_i is considered to constitute the power stroke of muscle contraction, in which the thick filament is pulled some 75 Å past the thin filament. In the final step, the S1 head is dissociated from actin by the binding of ATP, and the detached S1 unit returns to a perpendicular orientation near the thick filament. Finally, the hydrolysis of ATP by S1 sets the stage for the next catalytic cycle.

If this model of muscle contraction (which assumes that the power stroke is produced by a large scale structural change/swinging motion of the myosin head and that one stroke cycle corresponds to one ATPase catalytic cycle) were complete and generally applicable, at the monomolecular level it would require force fluctuations during the ATPase and power stroke cycles, with zero force in the ''off'' position (Figure 4–3). However, recent nano-manipulation studies of actomyosin motors *in vitro*—involving single-motor force recording comparable to single channel recording in membrane studies—find that 1:1 coupling between chemical and

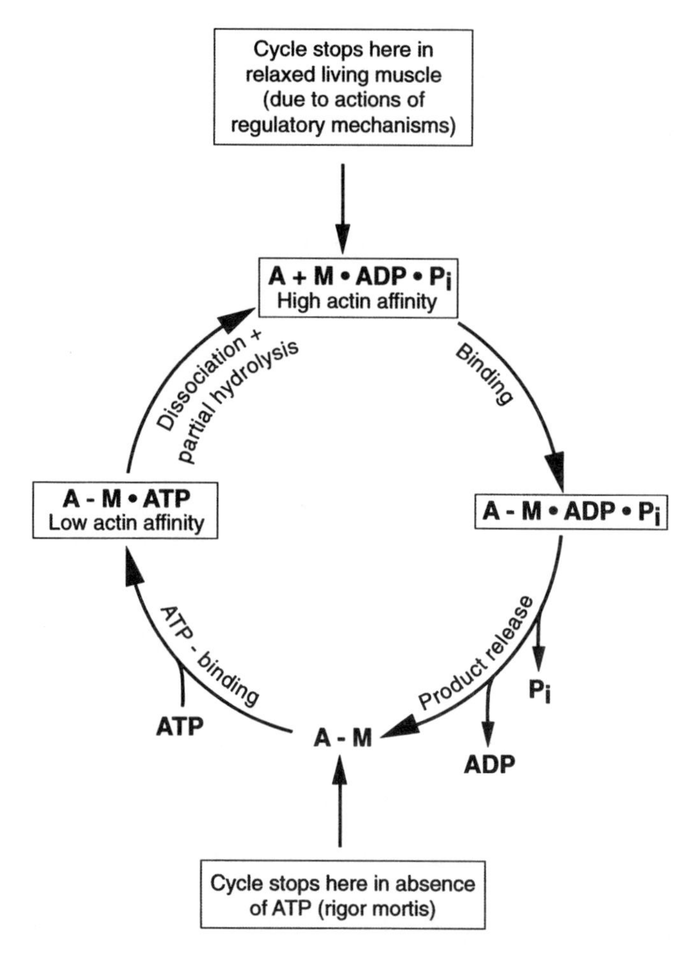

FIGURE 4–2. Four major steps in ATP hydrolysis by actomyosin when coupled with crossbridge cycling. Actin (A) in the thin filament and the myosin (M) crossbridge projecting from the thick filament interact cyclically. This interaction proceeds through a number of steps during which ATP is hydrolyzed, and the energy released ''drives'' or ''induces'' conformational changes in the crossbridge, allowing contraction. Each cycle causes the thick and thin filaments to interdigitate by about 10 nm. (Composite diagram assembled from several reviews [e.g., Ebashi, 1993].)

mechanical reactions and thus predicted large scale fluctuations in force over the ATPase and power stroke cycles appear to hold only at high work loads. At intermediate and low work loads, the actomyosin motor produces an almost constant force during the ATPase cycle. This indicates that the actomyosin motor at submaximal work loads can perform multiple power strokes during the ATPase cycle; i.e., the coupling between the ATPase and power stroke cycles is variable, depending on the load (Figure 4–4). If the free energy of ATP hydrolysis is fractionated into smaller packets available for several power strokes, the energy for each power stroke will necessarily be several times smaller than for the single overall ATPase reaction. How this is achieved is not understood at this time. Nevertheless, these new studies (Yanagida et al., 1993) emphasize that actomyosin is an exquisite molecular machine that can operate at high efficiency (up to and exceeding 40%), even when the input energy

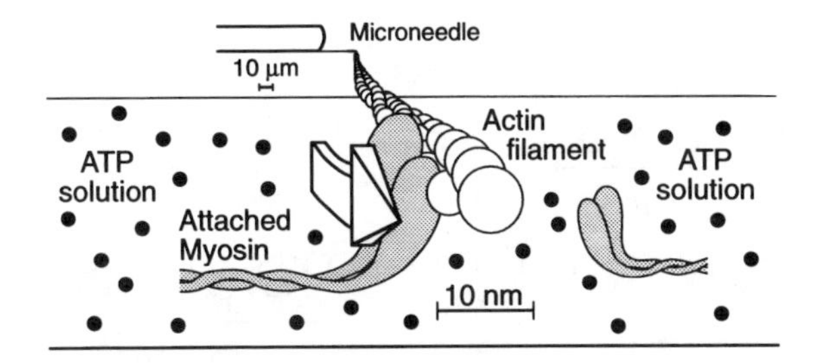

FIGURE 4–3. Reconstructing the muscle molecular motor *in vitro* requires a myosin-coated glass surface in an ATP solution. When labeled with fluorescent phalloidin, actin movement over the myosin can be quantified. Alternatively, one end of a single actin filament can be caught by a glass microneedle mounted on a micromanipulator and a piezo-actuator, while the other end is immersed into the ATP solution. The force generated on actin-myosin interaction can then be quantified by measuring the displacement of the microneedle in the direction shown by the arrow. Modified from Yanagida et al. (1993).

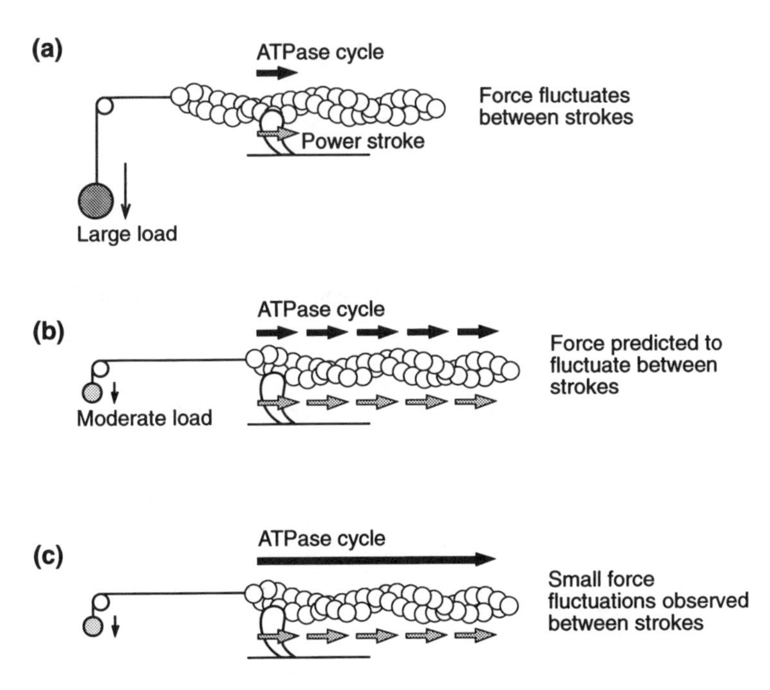

FIGURE 4–4. A new working principle of the actomyosin molecular motor, reviewed by Yanagida et al. (1993). (a) Under large load, near isometric conditions, tight 1:1 coupling is observed, with one power stroke (myosin head producing impulsive force) corresponding to one ATPase cycle and force fluctuating greatly between strokes. (b) Under moderate load conditions, the classical theory predicts 1:1 coupling and large force fluctuations, which are not experimentally observed. (c) Under moderate load conditions, the new model assumes variable chemo-mechanical coupling (1 ATPase cycle to many power strokes). In this model, the myosin head produces an almost constant force during the ATPase cycle and the force fluctuations are very small as a result; i.e., actin is moved a long distance smoothly and efficiently. Experimental observations are consistent with 1 ATPase cycle correlating with many power strokes under moderate loading conditions. Modified from Yanagida et al. (1993).

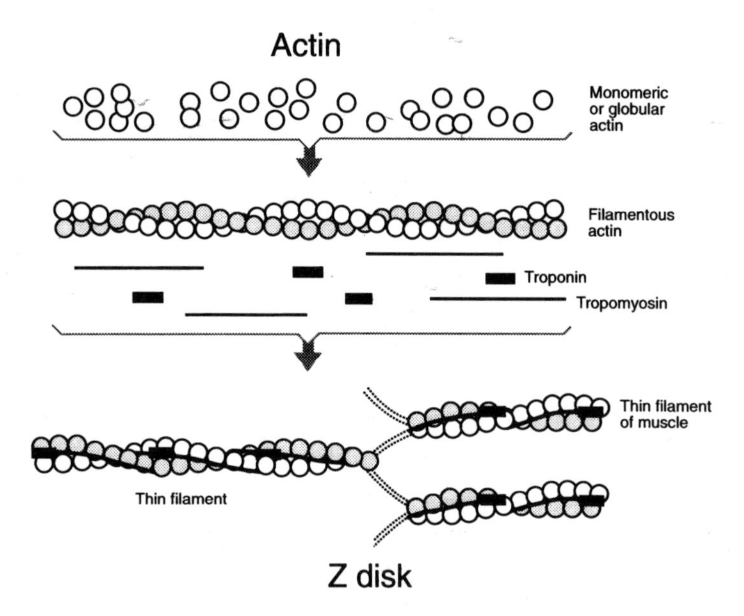

FIGURE 4–5. Composition and structure of the thin filaments in skeletal muscles. Globular actin monomers (top) polymerize into a two-stranded helical actin filament, which is completed by the addition a stiff, rod-shaped tropomyosin, and troponins which are bound to the tropomyosin. See text for further details. Modified from Murray and Weber (1974).

is close to the average thermal energy, in contrast to manmade machines, which operate at energy levels much higher than thermal noise. These exquisite functions of the actomyosin motor, at once a catalyst and a mechanical rachet, in the intact sarcomere are not possible without Ca^{++}.

TROPONIN AND TROPOMYOSIN MEDIATE Ca^{++} REGULATION OF MUSCLE CONTRACTION

Ebashi first discovered that the effect of Ca^{++} on the interaction of actin and myosin is mediated by tropomyosin and the troponin complex, which are located in the thin filament and constitute about a third of its mass (Figures 4–1 and 4–5). Tropomyosin is a two-stranded α-helical rod. This highly elongated 70-kDa protein is aligned nearly parallel to the long axis of the thin filament. Troponin (Tn) is a complex of three polypeptide chains: TnC (18 kDa), TnI (24 kDa), and TnT (37 kDa). TnC binds calcium ions, TnI binds to actin, and TnT binds to tropomyosin. The troponin complex is located in the thin filaments at intervals of 385 Å, a period set by the length of tropomyosin. A troponin complex bound to a molecule of tropomyosin regulates the activity of about seven actin monomers (Murray and Weber, 1974).

The interaction of actin and myosin is inhibited by troponin and tropomyosin in the absence of Ca^{++}, presumably because tropomyosin sterically blocks the S1 binding sites on actin units in an inhibited thin filament. Excitation of the sarcolemma and transverse tubules triggers the release of Ca^{++} by the sarcoplasmic reticulum. The released Ca^{++} binds to the TnC component of troponin and causes conformational changes that are transmitted to tropomyosin and then to actin. Specifically, tropomyosin is thought to move toward the center of the long helical groove of the thin filament, allowing the S1 heads of myosin molecules to interact with actin units of the thin filament. Contractile force is generated, and ATP is concomitantly hydrolyzed until Ca^{++} is removed and tropomyosin again blocks access of the S1 heads. Thus, Ca^{++} is thought to control muscle contraction by conformational adjustments in which the flow of information is

$$Ca^{++} \rightarrow \text{troponin} \rightarrow \text{tropomyosin} \rightarrow \text{actin} \rightarrow \text{myosin}$$

ADAPTABLE VS. CONSERVATIVE ASPECTS OF CONTRACTILE COMPONENTS

It is clear from the studies to this point that muscles in vertebrates are designed according to a universal sarcomere ''plan'', in which many structural and functional elements are strongly conserved (Huxley, 1985). Myosins and actins, for example, always interact along the same chemical pathway (involving the same interacting groups). Substrate binding and catalytic mechanisms in ATPase are also always the same. The lengths of *A*-bands in all vertebrate striated muscles are virtually identical (about 1.55 μm). The lateral spacings of the *A*-filaments (about 40 μm) hardly vary. In fact, according to Hugh Huxley (1985), evolution has pretty well maximized the packing of filaments both laterally and longitudinally (where spacing is limited by the minimum volume that a cross-bridge needs for function); in effect, muscles are as tightly and efficiently packed as possible, with individual cross-bridges being capable of generating more or less the same tension over each power stroke and hence of performing the same amount of work.

Yet, despite this impressive conservation of muscle design, we also know that muscles can be highly differentiated and finely tuned for specific, specialized kinds of tasks; indeed, we opened this chapter emphasizing such differences rather than emphasizing universal properties. Muscles, therefore, must also be reasonably plastic and adaptable. Before considering how, it is important to remind the reader that the above model of muscle contraction is widely but not universally accepted (see Pollack, 1984, for example). However, most discussion is about how the protein machinery works in converting chemical to mechanical energy; there is little argument about the actual components of the system. Our next goal is to consider how adjustable these components are during the adaptation of muscle for different kinds of functions. Perhaps the most basic mechanism for specializing muscle function involves the generation of isoforms of its structural components, so our analysis will start here.

Myosin Isoforms: Patterns and Distribution

As a result of several biochemical approaches (standard protein analysis, immunochemical specificity, plus cloning and genetic analysis) it is now evident that up to 13 genes are present for the heavy chains of myosins in mammals (Perry, 1985); at least ten are expressed in adult skeletal muscles (Pette and Staron, 1993). Current models propose that when skeletal muscle first forms in mammalian species, only an embryonic (or fetal) type of myosin gene is expressed, and only a fetal myosin isoform is observed. The induction of a slow myosin isoform can occur at any time in developing muscle fibers, providing the appropriate induction signals are present. If they are not, muscle fiber typically sustains transition to a neonatal myosin isoform (activated at birth) which, after a period of up to a few weeks, is in turn replaced by a fast myosin isoform (see Whalen, 1985, for literature). Thus, the overall development pathway can be summarized as follows:

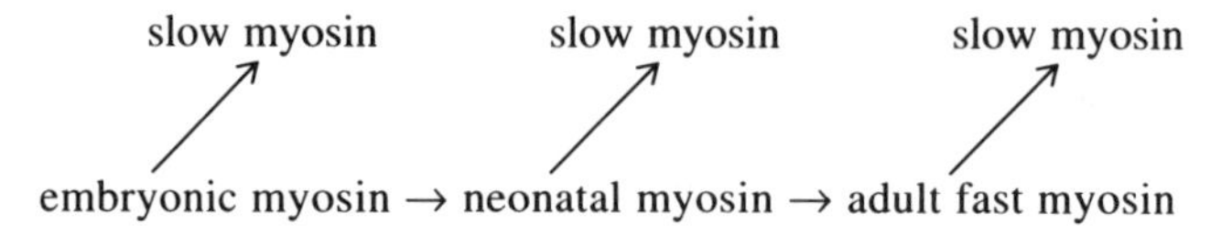

In this view, the differentiation of fast- and slow-type muscle fibers (or at least fast- and slow-type myosins) involves the specific activation of the gene for slow myosin only after at least one other myosin gene is already operative. Thyroid hormones may be involved (directly or indirectly) in slow myosin induction, although finer controls almost certainly are also required. Otherwise, it would be difficult to understand how the gene for slow myosin heavy chain could be turned on in cardiac muscle during fetal development but not turned on in skeletal muscle until the animal's adult-type muscle differentiation occurs (Whalen, 1985). Although the gross electrophoretic mobilities of each isomyosin triplet is determined by the specific heavy chain complement, the different mobilities of the isoforms within each triplet arise from different light chain combinations (Termin and Pette, 1991).

What are the functions of myosin isoforms? We think there are several kinds of functions for unique myosins in different fiber types. One possibility is that isoforms supply a mechanism by which to adapt interactions between thick and thin filaments, between myosin and ATP, or between myosin and its anchoring *M*-protein. As we shall see below, some of these processes have been investigated. However, another possibility (namely, that myosin isoforms may be a means for specific intracellular and intrafilament localization, as in *C. elegans* (Waterston and Francis, 1985), has not been considered in vertebrate systems and thus cannot be evaluated at this time.

Functionally, perhaps the commonest means by which the properties of muscles can be matched to their work requirements is through control of cross-bridge cycling rates, i.e. through control of myosin ATPase catalytic activity. Within a species, this turns out to be determined, in part at least, by the kind of myosin isozyme expressed (Reiser et al., 1985).

As far as we know, reaction kinetics and catalytic mechanisms for both slow and fast myosins are similar, with saturation occurring at 1 ATP per myosin head (i.e., 1 ATP per ATPase binding site). On theoretical grounds, the K_m for ATP should be in the n*M* range, and very careful measurements indeed indicate K_m values of about 30 n*M* or less (Hackney and Clark, 1985).

In contrast to ATP binding, the maximum catalytic capacity of fast myosin ATPase (expressed as V_{max} or turnover number) is about three times greater than for slow myosin ATPase. Both catalytic activities are activated about 200-fold or more by actin. In the presence of actin, the turnover number for fast myosin is about 10 s^{-1} compared to about 3 s^{-1} for slow myosin.

Another function for myosin isozymes may involve specializing myosin–actin interactions. While affinities for actin may be isoform-specific, this interaction is very selective to measurement conditions (such as ionic strength); at this time, no firm conclusions on this matter are possible (Woledge et al., 1985).

Actin Isoforms

Actin structure, in contrast to myosin, is not as adjustable, and actin is apparently the most strongly conserved of the muscle proteins; it is the only myofibrillar protein so far shown to be present in fast and slow muscles in identical forms. Isoforms are, however, present in smooth muscle, and even cardiac actin may differ from skeletal muscle actin (Perry, 1985). Conserved actin structure is expected because this protein in skeletal muscle must interact with a number of other proteins, which in contrast may occur in different isoforms. Selection for a strongly conserved actin structure may therefore be a single and direct outcome of needing to accommodate its interactions with the proteins interacting with it in the thin filament (Perry, 1985).

Tropomyosin Isoforms

In fast muscle, tropomyosin consists of two small subunits (each about 33,000 MW), termed α and β, and differing in 39 residues. The ratio of α:β in any given fast muscle appears very stable, with α subunits predominating. In slow muscle, the α subunit is absent or partially replaced with two additional forms, γ and δ, thus indicating that several isoforms may exist in a single muscle fiber. Although the functional roles of tropomyosin isoforms are not fully appreciated, their fiber-type specificity implies functional significance. Some evidence suggests, for example, that the α- and β-tropomyosin bind troponin T with differing affinities.

Troponin Isoforms

Different isoforms of troponins I, C, and T all have been found, but in many species, I and C tend to occur in a single form in a given fiber type. In contrast, troponin T, which occurs in three isoforms, may occur in more than one form per cell type. Unlike myosin isoforms, troponin T isoforms are generated by a novel mechanism of alternative mRNA splicing pathways from a single gene containing two different exons, which produce two different but related amino acid sequences near the C terminus of the protein (Hastings et al., 1985).

Troponin C in fast skeletal muscles is fiber-specific and contains two Ca^{++} binding sites (termed site I and II) per molecule. In interesting contrast, troponin C isoforms in slow muscle and cardiac muscle are identical and contain only one Ca^{++}-specific binding site, homologous to site II in TnC from fast muscles. Site I in this isoform contains amino acid substitutions, which abolish Ca^{++} binding. Since these sites must be filled for Ca^{++} to trigger contraction, it is evident that slow and cardiac muscles are activated by taking up 1 Ca^{++} per troponin C rather than 2 Ca^{++} per troponin C isoform in fast muscle.

Ca^{++} binding differences between slow and cardiac muscle troponin C may also occur, but these are ascribed to the differential activation of troponin C Ca^{++} binding by troponin I; the latter occurs in three isoforms, each specific to either fast, slow, or cardiac muscle, and may indirectly account for fiber-specific Ca^{++} binding properties (Perry, 1985).

Co-occurrence of Specific Contractile Isoforms

The above discussion suggests that while the contractile unit shows many standard features, its molecular composition shows a great deal of fiber-type specificity. Fast muscle fibers, for example, consist of (i) fast-type myosin (high ATPase), (ii) general muscle type actin, (iii) α- and β-type tropomyosin, and (iv) fast-type troponins I and T, and C, with the C-isoform displaying 2 Ca^{++} binding sites per troponin C molecule. Slow-type fibers contain (i) slow-type myosin (derived from gene activation only after either neonatal or fast myosin genes are turned off), (ii) general skeletal muscle type actin, (iii) β-, γ-, and δ-type tropomyosin, (iv) slow-type troponins I and T, and (v) slow-type troponin C (displaying only 1 Ca^{++} per troponin C molecule). In other words, the isoforms of the contractile machinery are linked in fiber-specific patterns, which correlate with functional specializations. Interestingly, this principle of co-occurrence of fiber-specific contractile components appears to extend to the excitation-contractile coupling phase in the overall process of muscle work.

EXCITATION-CONTRACTION (E-C) COUPLING: DECOUPLING OF Ca^{++} CHANNEL AND Ca^{++} PUMP FUNCTIONS OF THE SR

As already reviewed in Chapter 2, it is widely held that E-C coupling in skeletal muscle fibers of most animals involves the release of Ca^{++} from storage sites inside the fibre (mainly from the SR). The surface action potential is coupled to the SR, either directly at surface focal points or via the transverse tubules (T-system). The latter provide an electrical pathway from the surface to all portions of the fiber, basically acting as an inward electrical extension of the surface membrane. In frog fast-twitch fibers at room temperature, the surface action potential requires only a few milliseconds to spread throughout the matrix. As in surface action potentials, current through the T-system depends upon Na^+ and K+ fluxes in opposite directions (through Na^+ and K+ specific channels) leading to membrane depolarization. In the normal resting state, [Ca^{++}] in the cytosol is low (10^{-6} to $-10^{-8}M$) and this value is maintained through a 1:1 coupling between Sr Ca^{++} ATPase-mediated pumping of Ca^{++} into the cisternae and outward Ca^{++} fluxes. When the membrane voltage potential drops to some critical value, Ca^{++} channel and Ca^{++} pump functions become decoupled; Ca^{++} fluxes through Ca^{++} release channels momentarily exceed Ca^{++} ATPase pump fluxes, thus allowing a pulse of Ca^{++} to spill out into the cytosol. As discussed in Chapter 2, it appears that the TT Ca^{++} channels and the Ca^{++} release channels in the jSR occur as unique isoforms. Interestingly, there is also good evidence for such isoform occurrence of Ca^{++} ATPase, where both catalytic capacities (densities of Ca^{++} ATPase) and catalytic properties (isoform kind) vary with fiber type. Although Ca^{++} ATPases will be discussed in greater detail in Chapter 5, it is useful in this context to briefly review the situation.

Ca^{++} Pump Isoforms

Since the rate of channel-mediated Ca^{++} flux out of the SR in fast fibers is greater than in slow muscle, functional symmetry (in channel vs. Ca^{++} pump flux capacities) may require adjustments in the pump according to fiber type work demands. This principle in fact is observed, and it is now well documented that Ca^{++} uptake rates by fast muscle SR (Ca^{++} pump rates) exceed by some three-fold those values in slow muscle SR. This difference between Ca^{++} pumping capacity cannot be accounted for by differences in affinity of the Ca^{++} ATPase for Ca^{++}. Rather, it arises from higher Ca^{++} ATPase catalytic rates in fast muscle than in slow muscle. Not only does fast muscle sustain higher *amounts* of this catalytic capacity, it also appears to contain a fiber-specific isoform of the Ca^{++} pump. This is most clearly indicated by different regulatory properties; unlike fast muscle, in slow muscle, Ca^{++} ATPase is modulated via phosphorylation of phospholamban by cyclic AMP-dependent protein kinase. These features, plus analogous data on immunological reactivities, strongly suggest that different Ca^{++} pump isoforms are expressed in fast vs. slow muscle, a concept confirmed by molecular genetic studies showing two Ca^{++} ATPase genes, one unique to fast muscle, the second found in slow muscle and in the heart (MacLennan et al., 1985; 1986). In addition, slow muscle Ca^{++} ATPase pumps approximately 2 Ca^{++} per ATP hydrolyzed, a stoichiometry (and efficiency) nearly twice that of fast muscle ATPase (van Hardeveld, 1986). Thyroid hormones control the expression of both Ca^{++} ATPases at the level of mRNA production (Simonides et al., 1990).

CO-OCCURRENCE OF CONTRACTILE AND REGULATORY PROTEIN ISOFORMS

The framework or model emerging is that of an intriguing codevelopment of fiber-specific isoforms constituting the basic unit of skeletal muscle, the sarcomere, and its regulatory system. In effect, even if many molecular characteristics of muscle fibers are obviously potently conserved in overall structural outline, actual molecular details are clearly subject to close tuning of fibers designed for specific kinds of work.

We have already emphasized (see Chapter 3) that in muscles highly specialized for sustained speed, a lot of adjustment occurs at the level of information flow to the working muscle. In such muscles, the SR is greatly amplified, sometimes constituting up to 30% of the cell volume. For the same reasons, fast-twitch muscles have more SR per unit volume than do slow muscles, and all fast muscles contain relatively high Ca^{++} ATPase catalytic activities (Table 3–2). In all such cases, as far as we know, SR Ca^{++} release channels and Ca^{++} ATPases coadapt with specific myofibrillar isoforms structurally and with actomyosin ATPase functionally.

ROLE OF ACTOMYOSIN ATPase IN ADAPTATION OF MUSCLE FUNCTION

The functional significance of tuning up the actomyosin ATPase catalytic potential can best be illustrated by considering groups of homologous skeletal muscles (or groups of homologous activities).

Close inspection of available data indicates that, while overall sarcomere design is highly conserved, there is a wide range of ATPase reaction rates possible in muscles, both between and within individual species. In a benchmark paper nearly four decades ago, Hill (1950) pointed out that despite vast size differences, most animals differ very much less in the maximum speed at which they can move. He argued that this would be expected for animals of similar morphology with muscles capable of exerting similar tensions per unit area. It is the *velocity* of shortening of the muscle that varies most with animal size (being inversely proportional to the linear dimension of the muscle). Thus, the smaller the animal, the higher the velocity of sliding of actin filaments past myosin filaments, and the higher the actomyosin maximum ATPase catalytic potential (Table 4–1). Exactly the same considerations apply to fast vs. slow muscles within a given individual, where high and low actomyosin ATPase potentials codevelop with unique fast and slow isoforms of myofibrillar proteins. Insight into why this co-adaptation of parts is required may be obtained by analyzing the actomyosin ATPase function more closely. This analysis, parenthetically, will also help to explain the main paradox arising from the data in Table 4–1, namely, why estimates of *in vivo* ATP turnover rates sometimes vastly exceed accepted turnover rates of actomyosin ATPase measured *in vitro*.

IN SOLUTION, ACTOMYOSIN ATPases ARE HIGHLY ADAPTED CATALYSTS

To put this problem in perspective, we should re-emphasize that probably because of the coupling between the ATPase catalytic cycle and contraction, the turnover number for fast myosin is increased during actin association by a factor of about 200, while for the slow muscle isoform, the turnover number is about one-third this value. Recent studies indicate that each myosin head contains one ATP binding site and that ATP saturation curves show a sharp break when both sites are filled. Because of this behavior it has been difficult to obtain reliable K_m values for ATP. Modern techniques have at least partially resolved this problem and have reported a very low K_m value (about 30 n*M*), which is also predicted from rapid- or stop-flow kinetics and is considered a valid estimate of the affinity constant for fast myosins (Hackney and Clarke, 1985).

To our knowledge, the K_m for ATP for slow myosins has not been determined using similar advanced techniques. However, there is no reason to believe that it would be any higher than 30 n*M*; indeed, there may be reasons for assuming it would be lower.

Under conditions of limiting substrate and cofactor availability, the reaction velocity for complex reactions is determined by the parameter k_{cat}/K_m, where k_{cat} is the turnover number divided by the number of binding sites. For actomyosin ATPase, k_{cat} is one-half the turnover number (i.e., k_{cat} is the number of substrate molecules converted to product per unit time per active site when the enzyme is fully saturated). The ratio k_{cat}/K_m for fast actomyosin ATPase is $0.5 \times 10^1/3 \times 10^{-10} = 0.2 \times 10^9$ M^{-1} s^{-1} (Hackney and Clark, 1985). Because the K_m for ATP for slow actomyosin ATPase is as low or lower than that for the fast isoform, we assume that the k_{cat}/K_m ratio would be similarly high. Values of k_{cat}/K_m that fall between 10^8 to 10^9 M^{-1} s^{-1}, while not commonly observed, are known for several other enzymes as well. Such high ratios of k_{cat}/K_m show that such enzymes *can approach a kinetic ceiling.* When substrate levels at the active site are at or below the K_m value, *their catalytic velocity is restricted, primarily by the rate at which the active sites encounter substrate in solution.* Any further improvement of catalytic efficiency would be futile for such enzymes operating in solution, because their maximum reaction velocity would still be determined by diffusion-controlled encounter of the enzyme catalytic site with its substrate. When substrate concentrations are high, k_{cat} sets the rate ceiling for actomyosin, by release rates of products (ADP and Pi).

DIFFUSION LIMITATION OF ENZYME FUNCTION CAN BE CIRCUMVENTED

In metabolic regulation, it is possible to circumvent such substrate- or product-mediated limitation of enzyme function by effective ''channeling'' of reaction intermediates. That is, *for enzymes evolved into some form of kinetic limit, catalytic rates can be further enhanced only by doing something about the ceiling. In effect, what is needed is a delivery system for reactants that can work faster than diffusion.* Several means are known for achieving this end, but only two seem widely utilized in animals.

1. *Direct Handoff vs. The Fumble.* One of the most interesting possibilities for bypassing diffusion control of catalytic rate involves enzyme-enzyme direct transfer of metabolite (product of enzyme 1 is directly transferred as substrate to enzyme 2 without dissociation into solution). Direct enzyme-to-enzyme transfer of substrate is termed a ''handoff'', to contrast it with enzyme-to-enzyme transfers requiring dissociation into solution, a ''fumble''. For example, Srivastava and Bernhard (1986) supply evidence that GAPDH can directly transfer its products (DPG and NADH) to two target enzymes (PGK and LDH) without involving aqueous diffusion steps. Preferential access of substrate to the enzyme metabolizing it is a way of minimizing diffusion distance and falls into this category of ways for speeding up catalytic rates (see Ovadi (1991) and discussion forum following her review).
2. *Multienzyme Complexes as Direct Handoff Mechanisms.* Even more effective than the above mechanisms based on functional enzyme–enzyme interactions are multienzyme complexes in which enzymes in a given metabolic sequence

are tightly linked to form a stable multienzyme complex. In this case, channelling of substrate → intermediates → products is automatically assured with minimum possibility of dissociation into aqueous solution and diffusion limitation of the catalytic rate. PDH as a three-enzyme complex serves as an excellent illustration of this strategy. A point of emphasis is that once the substrate is channelled into a multienzyme complex, it is in effect locked there and sequestered away from all possible competing enzyme reactions. Thus *control* of the reaction pathway, as much as enhancing catalytic rate, is greatly facilitated. Indeed, facilitating control of metabolic reaction pathways is probably considered the main advantage of multienzyme complexes and may well explain why they arose in the first place.

THE CONTRACTILE CYCLE AS A CHANNELED REACTION SEQUENCE

Whereas in our discussion to this point it is evident that myosin ATPase per se is not directly covalently complexed with ATP-generating enzymes (such as CPK, PGK, or PK) of energy-yielding pathways, it is clearly intimately interacting with other proteins, and indeed the ATPase reaction is coupled to at least one other bonafide reaction or reaction sequence: cross-bridge cycling (the sequence: myosin-actin attachment, swivel, detachment). In a sense, actin may be viewed as a "substrate" for myosin ATPase where this enzyme-substrate complex (as in most metabolic reactions) is stabilized by weak or noncovalent bonding interactions. In this case, one "product" of the overall catalytic cycle is an actin binding site (n) moved relative to myosin through a distance of about 100 Å per catalytic cycle; simultaneously, the next actin "substrate" (the next actin binding site termed n+1) is moved into position for the next catalytic cycle (Figure 4–6). This is an example par excellence of vectorial reactions, which are common in metabolism. Whatever else one concludes, there is no doubt that myosin binding of actin is *not* dependent on diffusion, in which case the *selective advantages of direct "handoff-like" mechanisms for circumventing diffusion limitations for the adenylates and* P_i *moving into and out of the catalytic cycle may be very large indeed.* If this were so, how could it be shown?

In all studies of such mechanisms to this time, demonstration of a handoff requires proof that an enzyme-substrate complex serves as a competent substrate source for the target enzyme in the series. In the case of actomyosin ATPase, for example, one would need to know if the following reactions are favorable or possible:

CPK − ATP + H_2O binary complex at substrate concentrations (10^{-4} to 10^{-3} *M*)	→	actomyosin ATPase (catalytic quantities) $\sim 10^{-10}$ *M*	→	ADP + P_i
PGK − ATP + H_2O	→	actomyosin ATPase	→	ADP + P_i
PK − ATP + H_2O	→	actomyosin ATPase	→	ADP + P_i

By way of comparison, in the case of LDH, the analogous reaction, using the complex GAPDH-NADH at substrate-level concentrations as the source of NADH, is known to proceed faster than it does when diffusion-controlled (Srivastava and Bernard, 1986; Srivastava et al., 1989; Betts and Srivastava, 1991).

In summary, then, this analysis so far suggests that actomyosin ATPase *in vitro* displays such high kinetic adaptation that further increase in catalytic efficiency per

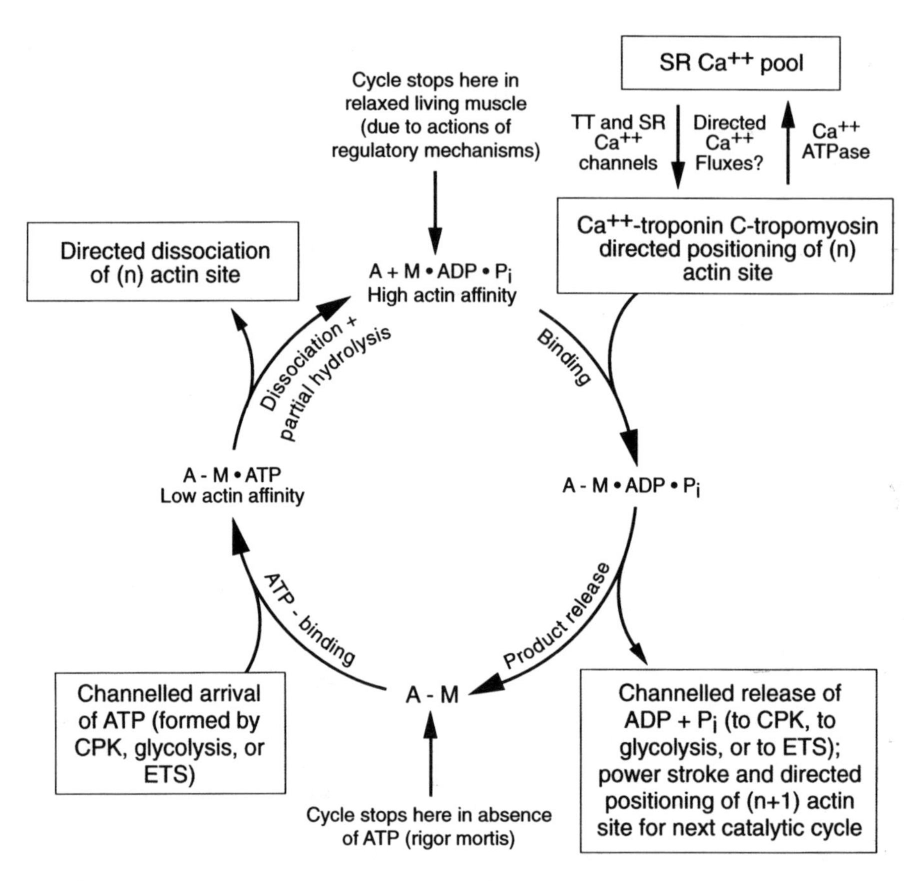

FIGURE 4–6. Reinterpretation of the classical model of major steps in ATP hydrolysis by actomyosin and its coupling to the crossbridge cycle. The diagram emphasizes processes that minimize dependence of the actomyosin motor on ATP, ADP, Pi, and Ca^{++} diffusion into and out of the ATPase cycle and the Ca^{++}-troponin-tropomyosin-regulated actomyosin power stroke.

se is relatively fruitless, since the reaction at limiting substrate already is about as fast as it can get in solution. Nevertheless, *in vivo ATPase turnover rates are higher than observable in vitro.* This paradox can be readily resolved if *in vivo* myosin ATPase functions as a channelled reaction with delivery of no required "substrates" or removal of no "products" being diffusion-limited; i.e., the paradox is solved by circumventing diffusional limitation of catalysis. Because of the structural requirements of *in vivo* function, one "substrate" for the myosin ATPase catalytic cycle (actin) is clearly not diffusion-limited. This part of the catalytic cycle is analogous to channelled reaction of multienzyme complexes such as pyruvate dehydrogenase (PDH) in standard metabolic pathways. Under these conditions, one would expect great selective advantage for effective channelling of the main metabolic substrate (ATP) and products (ADP + P_i) into and out of the catalytic cycle. The only demonstrated mechanism for this kind of channelling is termed the "handoff" and involves at least transient enzyme–enzyme interactions, allowing for improved substrate delivery rates. If this situation validly describes muscle function *in vivo,* myosin ATPase function would be analogous to a direct handoff of substrate to an enzyme complex such as PDH in mainstream metabolism, i.e., a combination of mechanisms (1) and (2) above.

EVIDENCE FOR PREFERENTIAL ACCESS TO AND FROM ACTOMYOSIN ATPase

As far as we know, preferential access mechanisms for adenylate and P_i participation in the myosin ATPase catalytic cycle are not firmly established. However, there are a number of instructive measurements available. For example, numerous studies show preferential binding of CPK and of some glycolytic enzymes to myofibrillar proteins. In the former case, preferential CPK binding is thought to supply ATP derived from this reaction with a preferential access to myosin ATPase, forming one arm of the PCr shuttle (Wallimann et al., 1992). Similarly, several glycolytic enzymes appear to display high affinities for myofibrils, and under some circumstances, large fractions of the total enzyme pool may occur in bound form. Aldolase, for example, displays this binding behavior, and its affinity for substrate (FBP) is increased nearly ten-fold when bound. This interaction may allow indirect coupling of myosin ATPase and glycolysis, while analogous PFK, PGK, or PK binding may allow the preferential exchanges of reaction intermediates for these metabolic steps as well (see Figure 1–1 and associated discussion).

Finally, it is worth emphasizing that, in all animals capable of extremely high aerobic muscle ATP turnover rates (small mammals, small birds, insects), there is extensive adjustment of structural organization of SR and mitochondria. The latter come to be very finely aligned with the myofibrils (and often with the storage substrate being burned), and mitochondrial volume density approaches 35% of cell volume (or approaching 1:1 with myofibril volume density). Typically, these muscle cells are also smaller than those commonly observed in vertebrates, which further minimizes diffusion distances. As in the case of CPK and glycolysis, what seems to be selected is the efficiency (or rate) of exchange of ATP, ADP, and P_i achievable between energy-yielding and energy-consuming structures.

The implications are rather profound: upward adaptations in muscle ATP turnover rate appear to be directed mainly at mechanisms for either completely circumventing diffusional limitation or at least for minimizing its impact (by minimizing diffusional distances). Adaptation in essence appears to be driving toward a structural organization where flux is controlled not by diffusional limitations but by the inherent *efficiency of interactions between metabolism and actomyosin ATPase, i.e., by the structure of the overall system:*

channeled arrival of *ATP* (formed by CPK, glycolysis, or ETS); directed dissociation of "n" *actin* site	ACTOMYOSIN ATPase →	channeled release of *products* (ADP and P_i) (to CPK, glycolysis or ETS) directed positioning of "n + 1" actin site for next catalytic cycle

MINIMIZING Ca^{++} DIFFUSION-BASED LIMITS

Exactly the same considerations as for the myosin ATPase catalytic cycle apply for fluxes of Ca^{++} (alternating between SR Ca^{++} channel mediated flux from SR → troponin C and Ca^{++} ATPase mediated flux back into the SR). For example, adaptations of SR structure in vertebrate fast, hummingbird flight, insect flight, and insect sonic muscles in effect minimize Ca^{++} diffusion distances from SR → troponin → Ca^{++} ATPase. As in the flux of adenylates into and out of the myosin ATPase catalytic cycle, there appears to be great selective pressure favoring mechanisms for minimizing Ca^{++} diffusion limitations (or for maximizing the possibilities for "directing" Ca^{++} traffic between specific sources and sinks). While structural design may not allow SR Ca^{++} to serve as a direct source of Ca^{++} for troponin C, for the reversed flux, Ca^{++} ATPase could have preferential access to Ca^{++} bound to troponin or to parvalbumin; thus, we need to know if Ca^{++}-troponin or Ca^{++}-parvalbumin are good substrates (i.e., good sources of Ca^{++}) for Ca^{++} ATPase. In this case, as in the contractile system per se, the ultrastructural evidence already available is consistent with mechanisms for minimizing diffusion dependence and clearly implies that natural selection is driving the system structure away from simple diffusional control of flux. Instead, the main determinants of flux that appear to be favored are the inherent *efficiencies of interaction and exchange (i) between SR Ca^{++} release channel and troponin C and (ii) between troponin-C bound Ca^{++} and Ca^{++} ATPase.*

5

Return to the Precontraction State

INTRODUCTION

Detailed discussion to this point has focused on signal transmission (motor neuron → muscle), culminating in Ca^{++}-activated muscle contraction. If enough time is available during relaxation, conditions at each of the steps in the above chain described are returned to the precontraction state. In overall outline, this requires the following initial series of steps. First, as already briefly described, $[Ca^{++}]$ in the cytosol is decreased back to resting levels by Ca^{++} ATPase-catalyzed pumping of Ca^{++} back into the SR. Second, $[Na^+]$ and $[K^+]$ in the cytosol are returned to normal levels by Na^+K^+ATPase-catalyzed pumping of K^+ back into the cell and Na^+ out of the cell. Third, at the synapse, ACh, which dissociates from end-plate channels, is hydrolyzed by AChE, forming acetate and choline in a reaction so rapid that it is considered unlikely that any given ACh would be bound by the end-plate channel twice. Fourth, choline, which is released in the reaction, is swept back into the presynaptic nerve cells by a high-capacity, high-affinity uptake system; once inside, it is recondensed with glucose-derived acetylCoA to form acetylcholine. The acetylcholine is then packaged back into vesicles in the SR and Golgi apparatus. Finally, $[Ca^{++}]$ in the cytosol of the nerve terminals is reduced to preactivation states by action of membrane-bound Ca^{++} ATPase.

The flow of information here cannot be viewed simply as a kind of reversal of the excitation path described above; precontraction states in general are attained by using different ion flux paths and different mechanisms. Since simultaneous function of both paths at once would be fruitless and energy dissipating, it is obvious that in normal, healthy muscle, mechanisms must exist for minimizing periods of overlapping function. In metabolic regulation, we would say that regulatory mechanisms keep such futile cycling to low levels. The implications are far reaching: for any given kind of muscle, the maximum frequency of contraction may be strongly influenced by the maximum rate at which these two paths (excitation and recovery) can flip back and forth, first one being favored and then the other.

We have already seen the component properties of the excitation-contraction part of the system; with overall design of the information flow systems for recovery now outlined, we next need to consider in greater detail the properties of its two dominant, ATP-consuming components (the Ca^{++} and Na^+K^+ pumps), as well as what criteria must be met in their design for specific kinds of muscles.

Ca^{++} ATPase AND SARCOPLASMIC RETICULUM (SR) STRUCTURE

As the Ca^{++} pump is considered to be located in the SR (and is a part of SR structure), a good way to start our analysis is with SR morphology, which is organized into two functionally distinct regions (see Catterall, 1991, for background literature).

One region is termed the terminal cisternae, lateral sacs, or junctional SR (jSR) and is found in close proximity to the transverse tubular system; a second region, the free SR (fSR), forms the longitudinal and fenestrated regions of the reticulum. Although the detailed structure may depend upon the muscle and animal type, this classification is consistent for all muscles, including the myocardium.

The close apposition of the jSR and the T-tubules results in the formation of triads. The flattened surfaces of two cisternal sacs of the jSR flank the junctional surface of a T-tubule. Occasionally, in the myocardium and the muscles of invertebrates, only one element of the SR forms a junction with a T-tubule (diad). The junctional gap between the T-tubule and the jSR is approximately 100 to 200 Å and is characterized by the presence of dense projections or feet, which connect the cytoplasmic leaflets of the two membranous systems. The feet are periodically dispersed at intervals of 300 Å and are arranged in rows of two or three on both sides of the T-tubule. The feet are considered to be associated with the jSR rather than the T-tubule, since they remain associated with the jSR upon fragmentation and fractionation of the reticular membrane.

Electron microscopy supplies us with one of the main clues for Ca^{++} ATPase localization on the fSR. Freeze-fractured fSR displays a dense population of 85 to 90 Å particles on the cytoplasmic fracture face. In contrast, its luminal face shows hardly discernable indentations and very few particles. The hypothesis that the intramembranous particles represent the Ca^{++} pump is supported by the correlation between the 85 Å particle density and (i) the rate of Ca^{++} influx, (ii) the Ca^{++}-stimulated ATPase activity, and (iii) the speed of contraction of various muscles. In fast-twitch muscle, the density of the 85 Å particles is *approximately 3,000 to 5,000/*μm^2 *across the entire surface of the fSR; in slow twitch muscle the particle density is approximately half that value.*

The fSR and the jSR differ strikingly morphologically and functionally. The 85 Å particles of the fSR are absent in the jSR region, but the appearance of larger and less densely packed particles is observed on the cytoplasmic leaflet after freeze fracture and immunoferritin labeling. The packing density of the jSR particles is much less than the fSR particles, and it is believed that the sharp transition between the two regions possibly represents a barrier to free intramembrane diffusion of protein molecules. The molecular basis for this barrier is not yet understood, but it emphasizes a separate structural basis for the Ca^{++} release channel vs. Ca^{++} pump functions of the SR.

Ca^{++} ATPase CATALYTIC CYCLE

How does the process of ATP hydrolysis drive the active transport of Ca^{++}? An important insight is provided by the observation that Ca^{++} ATPase is transiently phosphorylated in the overall catalytic cycle. The site of phosphorylation is the side chain of a specific aspartate residue, and the phosphorylated intermediate (E-P) is then hydrolyzed if Mg^{++} is present. A cycle of conformational changes driven by phosphorylation and dephosphorylation transports two Ca^{++} for each ATP hydrolyzed. The very high effectiveness of this ATPase enables it to transport Ca^{++} from the cytosol ($[Ca^{++}] < 10^{-6}M$) into the sarcoplasmic reticulum (where $[Ca^{++}] \cong 10^{-2}M$).

Can these kinetic data be rationalized with structural properties of Ca^{++} ATPases? The answer is probably affirmative, but to understand why, we must have a closer look at Ca^{++} ATPase structure.

MODEL OF Ca^{++} ATPase STRUCTURE

Perhaps the most important information arising from recent cloning studies of Ca^{++} ATPase is the complete base sequence of genes for Ca^{++} ATPases; predicted amino acid sequences are obtainable as a result. Armed with this information, MacLennan and his coworkers (1986) are attempting to rationalize kinetic data with a minimal model of Ca^{++} ATPase structure. According to this model, Ca^{++} ATPase contains several structural domains (each charged with a specific catalytic function). The main extramembranous domains are the nucleotide-binding domain (parallel beta sheet) hinged to the phosphorylation domain (parallel beta sheet) containing a key aspartate residue (Asp 351) and the transduction domain (antiparallel beta sheet). These domains are linked to the membrane by two pairs of helices that, together with S_1 from the amino terminus, form the Ca^{++} binding stalk. Ten helices (M_1 to M_{10}), in an arrangement yet to be determined, apparently form the transmembrane channel.

Additional structural features of special interest for the understanding of Ca^{++} translocation are the Ca^{++} binding sites in the amphipathic stalk helices. These are located at the boundary between the cytoplasm and the hydrophobic bilayer. These sites are of high affinity but are not of the ''E-F-hand'' configuration characteristic of high-affinity Ca^{++} binding to nontransport proteins. (This is reasonable since E-F-hand structures display constant affinity, whereas in the transport cycle, Ca^{++} is bound initially to sites of high affinity, which disappear after the enzyme is phosphorylated. Until the enzyme is dephosphorylated, only low-affinity Ca^{++} binding sites can be detected. These observations imply that the high-affinity Ca^{++} binding sites are dependent on protein configuration and that an oscillation between high and low affinities is achieved in catalysis through alterations of configurational states.)

According to MacLennan et al. (1986), the reaction cycle for Ca^{++} translocation begins with binding of two Ca^{++} and one ATP, each to high-affinity sites, respectively. This is followed (i) by a rapid, Ca^{++}-dependent phosphorylation of Asp-351 and (ii) by a step called occlusion, in which the rate of dissociation of bound Ca^{++} is decreased. The central reaction is an energy-transducing step in which two Ca^{++} are transferred from sites of high affinity to sites of low affinity in a process that is coupled to a decrease in the energy conserved in the aspartyl phosphate bond. The transducing step is followed by dissociation of Ca^{++} from internal low-affinity sites, dephosphorylation, and regain of the E_1 configuration.

An interesting feature of this model is that the Ca^{++} sites are made up from the amphipathic stalk helices, in which the high-affinity Ca^{++} binding sites are exposed to the cytoplasm while their hydrophobic faces are in the interior of the cluster of helices. Utilization of the energy conserved in the aspartyl phosphate bond in globular cytoplasmic domains allows the highly conserved sequences leading from these globular domains into the stalk sectors to rotate. At the level of the stalks, these rotations are thought to bring Ca^{++} into the interior of the cluster of helices and, at the same time, lower its binding affinity. As a result, *Ca^{++} is translocated into a ''cage'' that is not accessible to the cytoplasm but is accessible to a transmembrane channel.*

A consequence of the energetically unfavorable exposure of stalk faces to the cytoplasm is the creation of a force for driving the stalk sectors into the membrane. Support for this prediction of perpendicular movements of ATPase during the catalytic cycle comes from X-ray and neutron diffraction analyses of the relation of the mass of the ATPase molecule to the phospholipid bilayer during a single turnover induced by release of caged ATP that indicate that the enzyme becomes thinner and that about 8% of its mass moves into the membrane during the translocation cycle.

TWO Ca^{++} ATPase GENES: TWO (OR MORE) Ca^{++} ATPase ISOFORMS

Just as gene cloning studies indirectly yield critical insights into Ca^{++} ATPase structure and function by allowing deduction of amino acid sequence, so also do cloning studies yield important insights into Ca^{++} ATPase isoforms. In mammals, there are two genes for Ca^{++} ATPase. One Ca^{++} ATPase gene encodes for the enzyme in fast-twitch muscle; the second gene encodes for the enzyme in slow-twitch and in cardiac muscle. Thus, minimally two Ca^{++} ATPase isoforms are the rule, although alternative splicing of exons encoded in a single RNA is also possible, and thus, more than one kind of isoform per Ca^{++} ATPase gene is possible. Ca^{++} ATPase from fast-twitch muscle undergoes such an epigenetic processing during neonatal → adult maturation (MacLennan et al., 1986), as do other tissue-specific Ca^{++} ATPase isoforms (Maruyama et al., 1989).

TWO Ca^{++} ATPases: DIFFERENCES AND HOMOLOGIES

The amino acid sequence of slow and fast muscle Ca^{++} ATPases indicate 164 differences, 66 of which are conservative replacements; on the other hand, homologies are so substantial that the secondary structural topographies predicted for the two forms are very similar.

The amino acid substitutions appear in clusters that are related to the structural features of the molecule. Thus, most of the variation is found in α-helices and bends, the β-strands being highly conserved (12% variation). The variability of the helices depends markedly on their location in the molecule. The most variable are those at the amino terminus and in the phosphorylation and nucleotide-binding domains (34%). In contrast, the hydrophobic transmembrane helices (7.4%), the amphipathic helices of the stalk (7.6%), and the α-subdomain, or hinge domain (5%), are less variable. The α-subdomain, or hinge domain, and the stalk sector S4, and its extension into the phosphorylation domain, may conduct conformational changes between the cytoplasmic domains and domains directly involved in Ca^{++} binding and translocation. Interestingly, these two segments of the peptide chain, which join the parallel α-β sheet domains to the membrane, are the most highly conserved sequences in the molecule; this remains true when the comparison is extended to homologous ATPases that translocate Na^{+} and K^{+}, or H^{+} (McLennan et al., 1986).

These homologies suggest that the essential features of structure transmitting energy from cytoplasmic domains to transmembrane domains must be highly conserved. That is why it is rather surprising that the efficiency of the Ca^{++} pump (i.e., the amount of Ca^{++} transported per ATP hydrolyzed) varies in different types of muscles. In slow muscles, close to 2 Ca^{++}/ATP are pumped, which approximates the theoretical stoichiometry expected. In fast muscles, this can drop by up to 50%, implying either an uncoupling of Ca^{++} ATPase activity from Ca^{++} transport or some other form of slippage. The former may arise simply by selective modification of SR-membrane lipids, but no mechanism for slippage is known. Alternately, since these kinds of studies are often done with SR vesicles, the efficiency differences between slow and fast muscles could arise from more uncontrolled Ca^{++} efflux in fast than in slow SR (see van Hardenveld, 1986, for further discussion of this important topic).

COADAPTATION AND DESIGN PROPERTIES OF SR Ca^{++} ATPases

As Ca^{++} ATPase function can consume up to some 30% of the ATP turning over at high work rates, it is crucial that its properties be closely tuned to those of other components of the system structure. Although this area is sorely in need of more exploration, enough is known at least to indicate some of the highlight features. Minimally, these include:

1. SR Ca^{++} ATPase catalytic capacity must be reasonably closely correlated with SR Ca^{++} channel densities, for the faster the muscle, the faster the SR Ca^{++} release mechanism must operate and the faster the SR Ca^{++} ATPase must re-sequester cytosolic Ca^{++} during establishment of precontraction state conditions.
2. In general, therefore, we would expect and indeed find that Ca^{++} ATPases should occur in higher activities in fast than in slow muscles. At their upper limit, these kinds of upward adjustment can achieve over ten-fold increases in SR volume but only by about a two-fold maximum increase in Ca^{++} pump sites per square millimeter of SR membrane. There may be good reason for this; namely, that the SR membrane is already nearly maximally loaded with Ca^{++} ATPase molecules (constituting over 80% of the integral membrane protein and taking up about one-third of its surface area). Further adjustment in this direction may not be physically possible or may actually be counterproductive.
3. Along these lines, it is worth emphasizing that the Ca^{++} ATPase concentration in the SR membrane probably exceeds by far that of SR channels. This is because the relative catalytic efficiencies of the two differ by many orders of magnitude, channels inherently being kinetically far faster molecules than most ATP-dependent ion pumps. That is why, even if both channels and pumps obviously co-occur in the same general membrane, it is often found that fSR is so enriched in Ca^{++} ATPase that fSR purification is almost synonymous with purification of the enzyme! The fSR is often referred to as a semipurified Ca^{++} ATPase!
4. As mentioned earlier, many properties of muscles may depend upon the frequency with which the two information paths (activation/excitation/contraction vs. recovery) can flip back and forth. Thus, we may expect and indeed find that the requirements for regulation of Ca^{++} ATPases may show some fiber-type specificity. One such specificity, protein kinase regulation in slow-twitch muscle Ca^{++} ATPase, has already been discussed, with its sensitivity to phospholamban. Phospholamban, a phosphoprotein that regulates the activity of Ca^{++} ATPase, is localized in cardiac and slow-twitch fibers but is not found in fast-twitch fibers. In contrast, parvalbumin is preferentially localized in fast-twitch muscles and may play a role in speeding up Ca^{++} flux from troponin to Ca^{++} ATPase and thus back to the SR. During development, it appears that both SR and contractile protein gene families are controlled by a common myogenin induction factor and a common myogenic differentiation program (Arai et al., 1992).

$Na^{+}K^{+}$ ATPase AND THE Na^{+} PUMP

As outlined in Chapter 2, the action potential traveling across a muscle cell, leading ultimately to contraction, depends upon the concerted function of Na^{+} and K^{+} channels. Muscle cells, like most animal cells, have a high concentration of K^{+} and a low concentration of Na^{+} relative to the ECF. These ionic gradients are partially dissipated with each action potential and thus, particularly after hard work (Clausen, 1986; 1990), must be regenerated by a specific transport system that is called the *$Na^{+}K^{+}$ pump* because the movement of these ions is linked. The active transport of

Na^+ and K^+ is of great physiological significance, with more than a third of the ATP turnover of a resting animal being used to pump these ions. How is this achieved?

Since Skou first discovered a Na^+ and K^+ requiring ATPase in 1957 (see Post, 1989, for a brief history of this area) it has been hypothesized that the enzyme (termed Na^+K^+ ATPase) may be a biochemical representation of the Na^+K^+ pump, which *in situ* utilizes the energy released on ATP hydrolysis for actively transporting Na^+ and K^+. Evidence for this hypothesis comes from several kinds of observations:

1. Na^+K^+ ATPase occurs wherever Na^+ and K^+ are actively transported. Its catalytic capacity is quantitatively correlated with the extent of ion transport. For example, nerve cells are rich in both the Na^+K^+ ATPase and the pump, whereas muscle cells have low levels of both.
2. Na^+K^+ ATPase and the pump both are tightly associated with the plasma membrane in all cells.
3. Na^+K^+ ATPase and the pump both are oriented in the same way in the plasma membrane in all cells.
4. Variations in $[Na^+]$ and $[K^+]$ influence ATPase activity and the rate of transport of these ions in parallel ways.
5. Na^+K^+ ATPase and the pump both are specifically inhibited by cardiotonic steroids; what is more, the K_i (concentration of inhibitor causing half-maximal inhibition) is the same for both processes.
6. The pump can be reversed under suitable ionic conditions so that ATP is synthesized from ADP and P_i.

Na^+K^+ ATPase CATALYTIC CYCLE

In considering how Na^+K^+ ATPase works, it may be worth emphasizing that there are many similarities between Ca^{++} ATPase and Na^+K^+ ATPase. In particular, during both catalytic cycles, both enzymes are phosphorylated; in both, an aspartate residue is the site of phosphorylation, and in both, conformational changes are the basis of ion translocation. In the case of Na^+K^+ ATPase, the phosphorylated intermediate (E-P) is hydrolyzed if K^+ is present; the phosphorylation does not require K^+, whereas the dephosphorylation reaction does not require Na^+ or Mg^{2+}. The pump also interconverts between at least two different conformations, E1 and E2. As a result, at least four states (E1, E1-P, E2-P, and E2) are involved in the transport of these ions with concomitant ATP hydrolysis. The stoichiometry is 3 Na^+ and 2 K^+ per ATP hydrolyzed, which is why the pump generates an electric current and a potential across the plasmalemma. The maximal turnover number of the ATPase is about 100 s^{-1}. As in the case of the Ca^{++} pump, a key feature of the Na^+ pump is that ATP is not hydrolyzed unless the two ions are translocated; i.e., the system is coupled, and the energy stored in ATP is not dissipated unless ions are pumped.

Na^+ K^+ ATPase STRUCTURE

Na^+ K^+ ATPase is a 270 kDa $\alpha2\beta2$ tetramer. A part of the α subunit (95 kDa) forms the ATP binding site, as well as a site that binds cardiac glycoside inhibitors, while the smaller β subunit contains carbohydrate residues. Several lines of evidence suggest that the ATP catalytic site is on the cytosol side of the membrane, while the steroid-inhibitor site is on the extracellular side of plasmalemma (Figure 5–1). Molecular biological studies have shown that the α subunit consists of 1016 amino acids

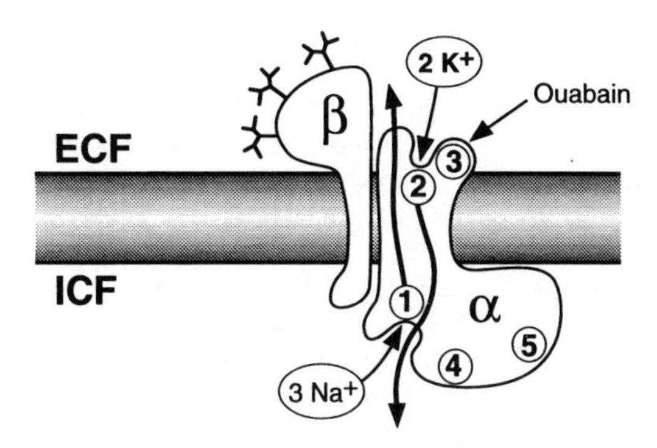

FIGURE 5–1. Current hypothetical model of subunit components of Na^+K^+ ATPase. Glycosylation sites shown on β-subunit. Na^+ channel (1) and K^+ channel (2) functions are coupled in a 3:2 stoichiometry. Domains (4) and (5) refer to potential phosphorylation sites. Domain (3) includes the ouabain binding site. Modified from Mercer (1993).

with 8 transmembrane domains, an intracellularly-located phosphorylation site (aspartate 369) and likely candidate regions for other pump-specific functions (Shull and coworkers, 1985).

MINIMAL DESIGN CRITERIA FOR Na^+ K^+ ATPase AS AN ION PUMP

Although the details of how phosphorylation and dephosphorylation result in ion translocation are still being sorted out, many workers in this field consider that any model of the protein pump minimally must fulfill three basic criteria (Mercer, 1993; Post, 1989):

1. It must contain a channel or cage large enough to admit ions,
2. It must be bicameral, so that the cage can be open to the inside in one conformation and to the exterior in another, and
3. The affinity for the transported ions necessarily must be different in the two configurations.

Na^+ PUMP ISOFORMS BASED ON α- AND β-SUBUNIT POLYMORPHISM

In view of the ubiquity of the Na^+ pump requirement, it is not surprising that the macromolecule responsible for performing it should be highly conserved in structure between different tissues and species. Nevertheless, it is now known that at least three genes and thus three α subunit isoforms occur in mammals and birds α-1, α-2, and α-3. The first predominates in the kidney, while α-2 and α-3 are predominant in excitable tissues, including skeletal muscle, heart, and brain. Overall tissue and cell specificity is complex and development stage-specific. Some regions of this subunit are highly conserved in evolution and are found essentially universally; other regions are more restricted to specific phylogenie and are seemingly less stringently—or at least differently—monitored by natural selection (Pressley, 1992).

At first, it was considered that only one gene specified β-subunits, but recently, evidence for at least three β-subunit isoforms has accumulated. The β-1 isoform is found in muscle and most other tissues, while β-2 is restricted to nervous tissue; β-3 is developmentally regulated and also brain-specific (Mercer, 1993). How these interact with the α subunits is not known.

FUNCTIONAL SIGNIFICANCE OF Na^+ PUMP ISOFORMS

Although the different isoforms of Na^+K^+ ATPase are apparently products of different genes, most of their functional properties (substrate and ion affinities, site–site interactions, and ion transport stoichiometries) are quite similar (Horisberger et al., 1991). Many workers have therefore pondered the significance of these isoforms. There are several possibilities.

First, Na^+ pump isoforms could play a role in metabolic regulation. For example, in skeletal muscle (and adipose tissue), insulin action depends upon Na^+K^+ ATPase activation; apparent only α-2 or α-3 isoforms account for this response.

Second, isoforms could contribute to Na^+ pump regulation per se. Recent studies (Carilli et al., 1985) indicate that endogenous inhibitory factors act on the extracellular side of the membrane by mechanisms similar to that of ouabain and other cardiotonic steroids. The Na^+K^+ ATPase isoforms display up to 1000-fold differences in sensitivity to ouabain; thus, the sensitivity of a given cell type to endo-ouabains presumably would be strongly influenced by the ratio of the different isoforms expressed in that tissue. The α-2 and α-3 isoforms present in muscle are the most sensitive to ouabain or other cardiotomic steroids, while α-3 is least sensitive.

Third, isoforms could play a role in tissue-specific cell localization or in protein–protein interactions. Numerous examples in the literature report specific enzyme–Na^+ pump interactions; one of the most intriguing is that of a reasonably tight interaction between CPK and Na^+K^+ ATPase. At least in one tissue, noninvasive *in vivo* NMR spectroscopic measurements are suggestive of a strongly preferential access of CPK-generated ATP to the Na^+K^+ ATPase (Blum et al., 1991). Such roles in overall metabolic integration may help to explain the co-occurrence of α2 Na^+K^+ ATPase in cardiac hypertrophy, along with the reappearance of a unique myosin isoform—both correlating with a decrease in speed of shortening of muscle fibers and with improved efficiency of muscle work (Charlemagne et al., 1986).

The theme from our earlier analyses may well be emerging again: specific isoform batteries are used to build specific kinds of muscle fibers. However, for the Na^+K^+ ATPase isoforms, much further work is clearly required to clarify isoform–isoform relationships and interactions.

MAGNITUDE OF THE POSTEXERCISE Na^+K^+ IMBALANCE

Until recently, the quantitative significance of the problem of postexercise Na^+ K^+ imbalance was overlooked because most sampling sites were in venous blood-draining resting muscle, which clears the plasma of extra K^+ by the action of the Na^+K^+ ATPase. What exposed the magnitude of the problem were recent measurements of very high $[K^+]$ in the ECF and in arterial blood. For example, if we consider the rat soleus, it has a total capacity for K^+ accumulation via the Na^+K^+ ATPase of about 6200 μmol g muscle^{-1} min,$^{-1}$ while the efflux of K^+ is about 7 μmol g^{-1} min^{-1}. At an excitation frequency of 15 Hz, the efflux of K^+ comes to $15 \times 60 \times 7 = 6300$ μmol g^{-1} min−1. Thus even at this moderate contractile activity, the muscle sustains a net loss of K^+. Numerous literature observations are consistent with these calculations, indicating the need for regulation of Na^+K^+ ATPase for correcting the K^+ imbalance during recovery.

SHORT-TERM Na^+K^+ ATPase REGULATORY MECHANISMS

In muscle, the most common moment-to-moment mechanism for regaining Na^+ and K^+ homeostasis is the response of the Na^+K^+ ATPase to changes in substrate concentrations. The combined effects of increased intracellular [Na^+] and extracellular [K^+] promptly activate the pump and so ensure that their own concentrations are returned toward the normal range. Additional to this fundamental kinetic regulation of ionic homeostasis are the effects of work per se and of hormones. Electrical stimulation produces potent (voltage-dependent) activation of the Na^+K^+ pump (and of ion translocation) before there is any measurable change in ionic concentrations, and this effect seems to be larger in slow than in fast muscles (Clausen, 1990). This regulatory effect is normally followed with up to a 100% catecholamine-mediated stimulation of Na^+K^+ pump activity. Catecholamine regulation occurs within a few minutes of exercise, leads to a redistribution of Na^+ and K^+, and occurs in both animals and humans. Insulin likewise activates the Na^+K^+ ATPase, again in both animal models and in humans, although less information is available on whether this mechanism is used in routine muscle activity (Clausen, 1990).

LONG-TERM Na^+K^+ ATPase REGULATORY MECHANISMS

All of the above regulatory mechanisms rely basically on the kinetic control of the Na^+K^+ pump; they are immediate (which is their considerable advantage), but they are limited by the maximum catalytic activity expressed at any given time in the target tissue. Longer term regulatory mechanisms adjust this ceiling by up- or down-regulation of the Na^+K^+ ATPase numbers or densities. The main mechanism for achieving this is through thyroid hormone (T3) regulation of Na^+K^+ ATPase biosynthesis. Four lines of evidence for this kind of biochemical adaptation are particularly convincing, namely, studies of training effects in animals and man, of reduced caloric intake in animals, of K^+ deficiency effects in animals, and of chronic diuretic effects in man. Each of these reports changes in enzyme amount as the main means for controlling the Na^+K^+ ATPase activity (Clausen, 1990).

Even longer term adaptations—those involving phylogenetic time and punctuating differences between species—are also clearly known. At least in the vertebrates, it appears that these interspecies adaptive differences in Na^+K^+ ATPase content in homologous tissues are also probably mediated mainly by thyroid hormones (Hulbert, 1987).

MUSCLE Na^+K^+ ATPase FUNCTIONAL DESIGN CONSIDERATIONS

Many of the functional constraints discussed with regard to Ca^{++} ATPase apply equally to the Na^+K^+ ATPase. However, this pump is not as crucial in moment-to-moment regulation of muscle function as is the Ca^{++} ATPase; thus its concentration in muscles is comparatively modest. Although Na^+K^+ ATPase densities vary in different muscles, within those species that have been examined, densities differ by only three- to four-fold range and are in all cases a long way from theoretical maximum packing densities. It might be noted, however, that the fastest muscles known—in insects or vertebrates—have not been examined in this regard. If we assume that the fraction of basal metabolism ascribable to the Na^+K^+ pump is reasonably constant between species (say 20 to 40% of BMR), then even in the fastest muscles, it is unlikely that the Na^+K^+ ATPase densities would ever approach theoretical maximum packing densities; the need is simply not high enough. (Only in tissues specialized for ion transport, such as gills or kidney tubules, is this ceiling approached.)

A more serious constraint on Na^+K^+ ATPases in different kinds of skeletal muscles is that of matching pump densities with the densities of channels, transporters, and exchangers with which their functions must be integrated. The problem arises, as already discussed for Ca^{++} ATPase integrated function, because of very different catalytic efficiencies of ion channels and ion pumps. Hence, the impact of any up- or down-regulation of either kind of component must be assessed with reference to its relative catalytic efficiency. The magnitude of this constraint can be illustrated by considering disease states in which Na^+ channel and Na^+K^+ ATPase densities are independently changed. In myotonic dystrophy (Steinert's disease), Na^+ channel densities in muscle remain normal, while Na^+K^+ ATPase densities drop to as low as one-sixth of normal. In a neurogenic muscular atrophy disease termed "lower motor nerve impairment," Na^+ channel and Na^+K^+ ATPase densities both rise by a factor of two. Because channels are perhaps 1000 to 10,000 times more catalytically efficient than pumps, decreasing Na^+ pump densities to one-sixth of normal represents an enormous impact on maximum absolute fluxes through these two proteins. Similarly, in lower motor nerve impairment a two-fold rise in Na^+ channel densities represents a potential increase in Na^+ channel fluxes up to 10,000 times greater than the potential increases in Na^+ pump fluxes caused by a two-fold increase in Na^+K^+ ATPase densities. Little wonder these muscles display electrophysiologically based functional abnormalities! Similar studies of Na^+K^+ ATPase and Ca^{++} ATPase densities in denervated human muscles underline how critically important it is for change in one part of the system structure to be balanced by appropriate change in other parts (see Hochachka, 1988a, for more extensive analysis of this area). From our perspective at this time, these studies emphasize an important design requirement for such coadaptations, namely, that a fractional change in one part of the system (e.g., channel densities) may require enormous changes in coordinated parts (e.g., pump densities) because of enormous differences in the catalytic efficiencies of the different components forming the system structure.

6

Supplying Muscle Machines with Energy

INTRODUCTION

To this point in our analysis of muscle machines, we have only looked at the ATP-consuming side of the effector system. But on endogenous ATP supplies alone, muscle work, of course, would quickly grind to a halt. That is why a complete description of muscle machines also requires analysis of ATP-generating mechanisms to account for how energy utilization by working muscles can be matched by energy production rates.

Metabolic pathways producing ATP as a utilizable energy source are of two types, anaerobic (or O_2-independent) and aerobic (or O_2-dependent). It is necessary for us at this point to briefly overview the main pathways utilized by muscle tissues.

THREE BASIC ATP-SYNTHESIZING PATHWAYS IN MUSCLE

The simplest mechanism for generating ATP is phosphagen mobilization. In vertebrate tissues such as muscle containing creatine phosphate (PCr), this mobilization is catalyzed by creatine phosphokinase (CPK), a process that requires no O_2 and can be written as follows:

$$PCr + ADP + H^+ \rightleftharpoons ATP + creatine$$

Fermentation, or the partial (O_2-independent) catabolism of substrates to anaerobic end products, is a second means of forming ATP. In animals, the most common fermentative pathway is that of anaerobic glycolysis (Figure 6–1). At high pH (>8.0), the summed reaction can be written as follows:

$$glucose + 2ADP^{3-} + 2HPO_4^{2-} \rightarrow 2\ lactate^- + HATP^{4-}$$

At low pH (<6.0), the adenylates are more protonated and the reaction can be written:

$$glucose + 2HADP^{2-} + 2H_2PO_4^- \rightarrow 2\ lactate^- + 2H^+ + 2HATP^{3-}$$

The pathway of glycolysis is phylogenetically ancient, and many of its features are strongly conserved (Boiteux and Hess, 1981). However, the terminal dehydrogenases are subject to adaptive change; in many invertebrate organisms, lactate dehydrogenase is replaced by functionally analogous imino acid dehydrogenases. In such an event, an imino acid (octopine, alanopine, strombine, or tauropine) replaces lactate as the anaerobic glycolytic end product. However, neither the proton stoichiometries nor the ATP yields are changed (Hochachka and Mommsen, 1983). If glycogen is

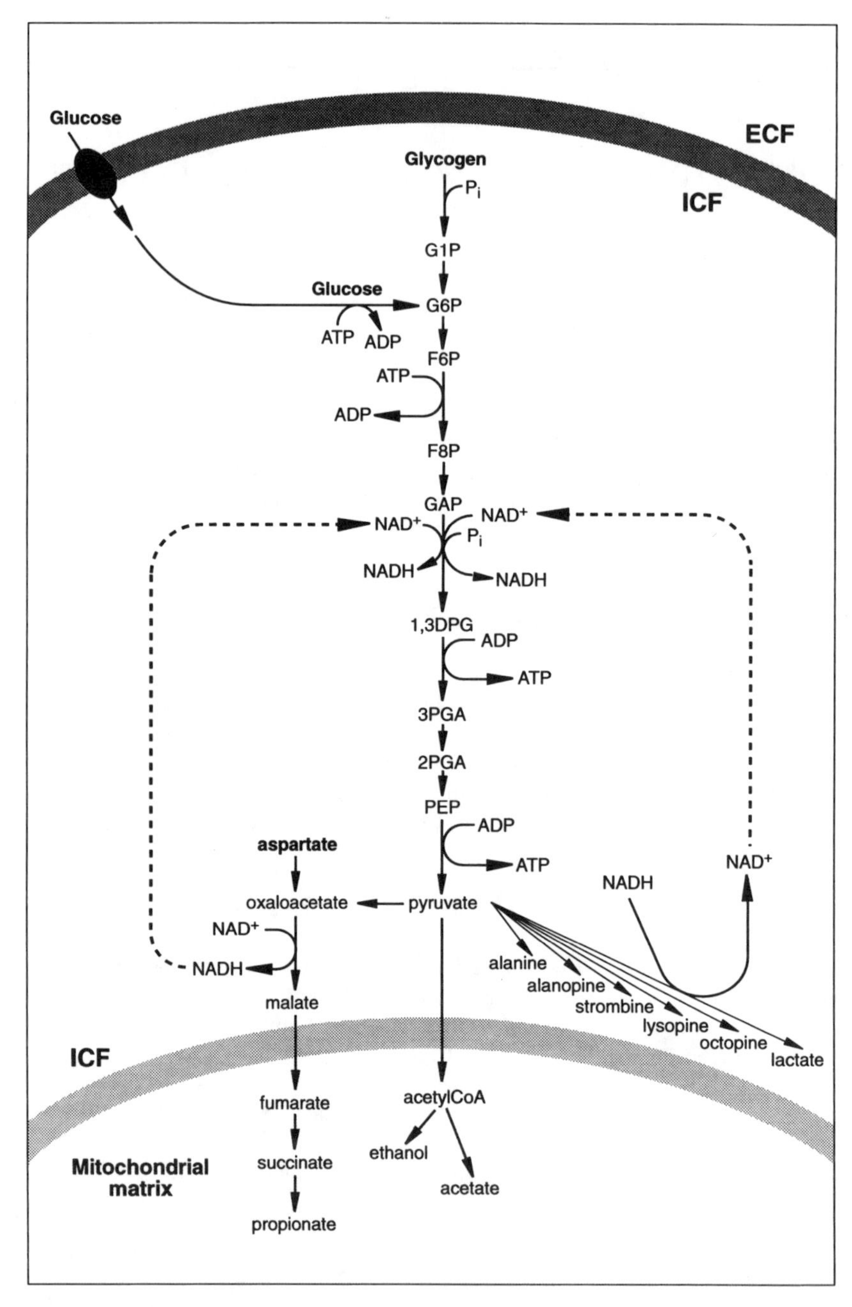

FIGURE 6–1. Major pathways of anaerobic metabolism in muscles of vertebrates and invertebrates. In most vertebrates, the dominant anaerobic pathway in muscle tissue is anaerobic glycogenolysis (glycogen → lactate) or anaerobic glycolysis (glucose → lactate).

the substrate fermented instead of glucose, the yield of ATP is 3 moles ATP formed per mole glucosyl unit.

In addition to carbohydrates, some amino acids can also be fermented. Aspartate, for example, can be fermented to succinate or propionate, a process that in mollusks is stoichiometrically coupled to glucose (or glycogen) fermentation. A third major category of potential fuel, the fats, is so reduced that they are not fermentable by animal cells.

In animal fermentations, an organic molecule (e.g., pyruvate) serves as a terminal proton and electron acceptor, forming an organic end product (e.g., lactate). In contrast, O_2 is required as a terminal acceptor for the complete oxidation of substrates such as glucose, glycogen, fatty acids, or amino acids. The pathways by which such complete oxidations are achieved are much more complex than most fermentation pathways. In the case of glucose, the first phase of complete oxidation (namely, the conversion of glucose to pyruvate) is the same as in glycolysis. However, instead of being reduced to lactate, pyruvate is converted to acetylCoA, which serves as the entry substrate into the Krebs cycle, a process occurring in the mitochondrial matrix (Figure 6–2). The two carbons of acetylCoA appear as CO_2 with the simultaneous formation of reducing equivalents in the form of NADH. Under anaerobic conditions, NADH is reoxidized to NAD^+ by an organic substrate that is reduced in the process (pyruvate, for example, being reduced to lactate; fumarate similarly can be reduced to succinate). Under aerobic conditions, NADH is reoxidized to NAD^+ via the electron transport system (ETS) located in the mitochondrial inner membrane or cristae (Figure 6–3). During the electron transfer process, H^+ ions are believed to be pumped out of the mitochondria across the inner mitochondrial membrane.

The developed H^+ ion concentration gradient and an electric potential across the membrane are thought to supply the driving force for ATP synthesis from ADP and P_i, a thermodynamically unfavorable reaction catalyzed by ATP synthase. The latter is a mitochondrial enzyme located on and spanning the inner mitochondrial membrane. At least when in submitochondrial particles, ATP synthase saturation kinetics involve ADP-positive site–site interactions in catalysis.

The net reaction for glucose oxidation (via glycolysis, the Krebs cycle, the ETS, and ATP synthase) can be written as follows:

$$\text{glucose} + 36(\text{ADP} + \text{P}_i) + 6\text{O}_2 \rightarrow 36\text{ATP} + 6\text{CO}_2 + 42\text{H}_2\text{O}$$

When fatty acids are the fuel being combusted, the pathway of oxidation is termed the β-oxidation spiral which, following fatty acylCoA formation involves four basic steps:

1. acylCoA + FAD^+ → α-β-unsaturated acylCoA + $FADH_2$
2. α-β-unsaturated acylCoA + H_2O → β-hydroxyacylCoA
3. β-hydroxyacylCoA + NAD^+ → β-ketoacylCoA + NADH + H^+
4. β-ketoacylCoA + CoASH → acylCoA (−2 carbons) + acetylCoA

This pathway, also located in the mitochondria, generates acetylCoA, which is then completely metabolized by the Krebs cycle, the ETS, and ATP synthase. The overall equation for fatty acid oxidation, using palmitate as an example, can be written as:

$$\text{palmitate} + 23\text{O}_2 + 129(\text{ADP} + \text{P}_i) \rightarrow 129\text{ATP} + 16\text{CO}_2 + 145\text{H}_2\text{O}$$

In most mammalian muscles, amino acids are not a major fuel for oxidative metabolism, although they may be in numerous other animals. When amino acids are

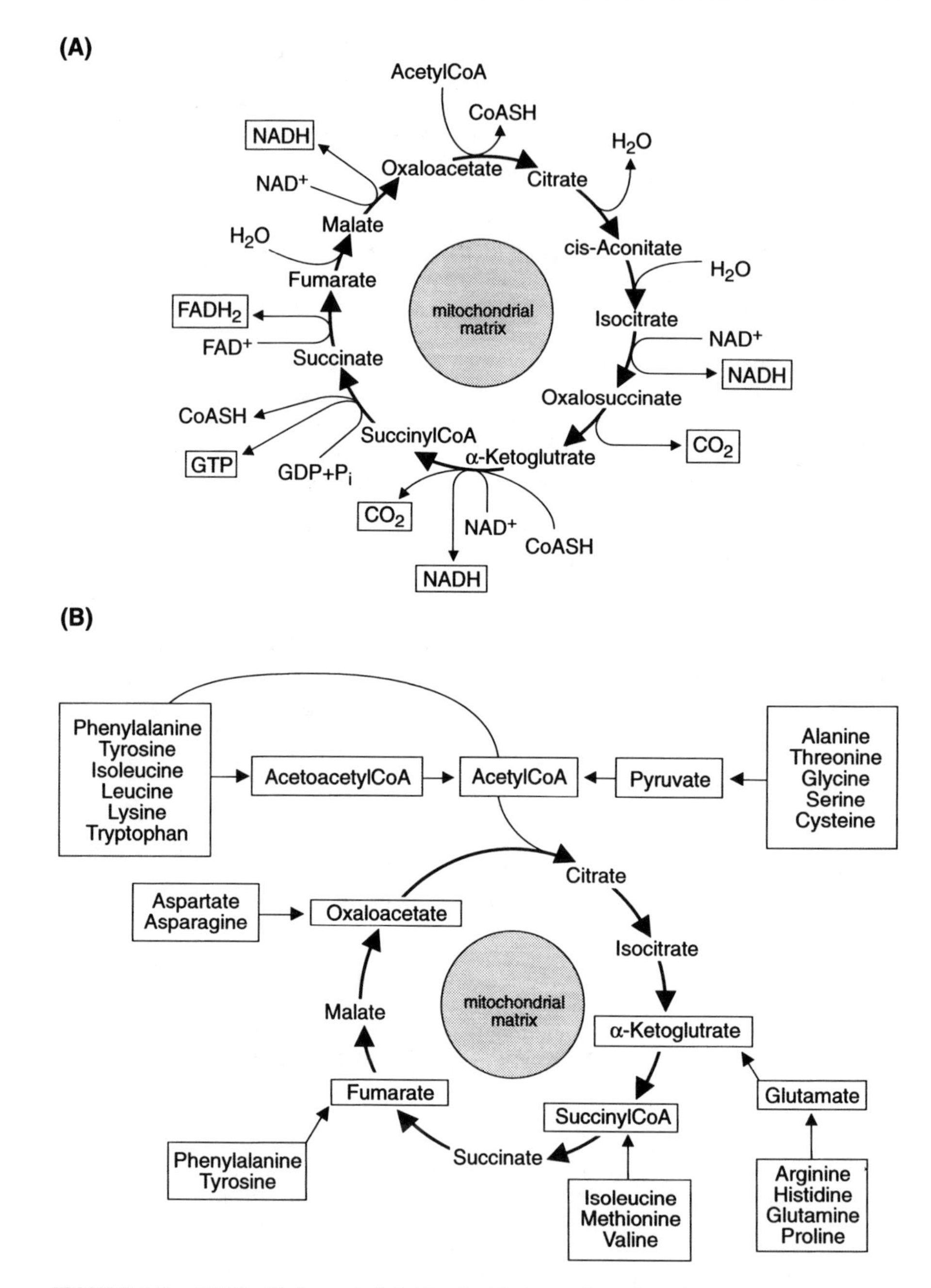

FIGURE 6–2. (A) The Krebs cycle is initiated with the condensation of oxaloacetate and acetyl Coenzyme A and ends with the formation of oxaloacetate from malate; this is considered to be a major final and common pathway for the complete catabolism of most carbon fuels. (B) Sites of entry of amino acids into the Krebs cycle are requisite steps for the complete catabolism of amino acids and proteins.

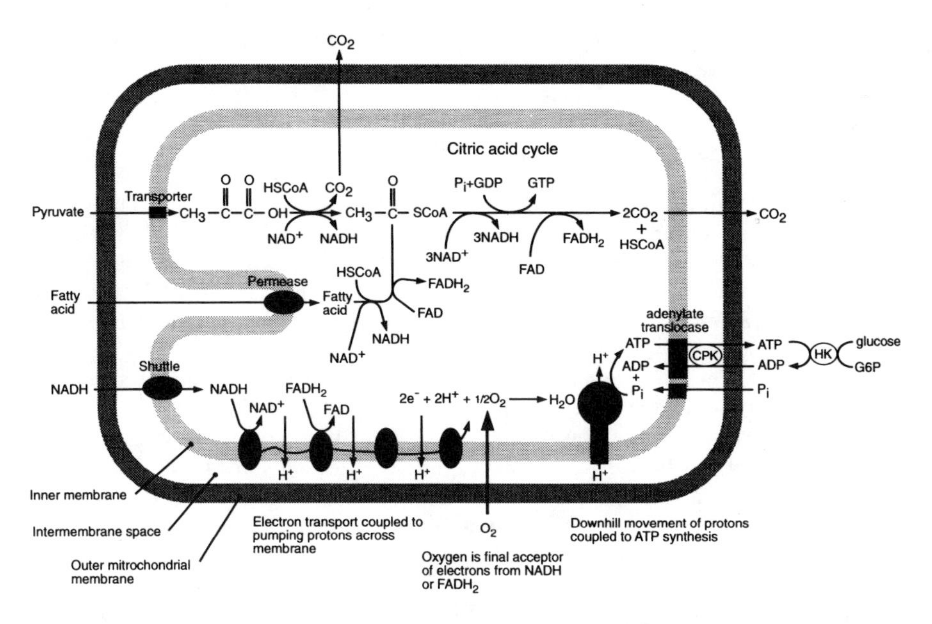

FIGURE 6–3. An abbreviated summary of the main processes associated with ATP synthesis in mitochondria. The substrates of oxidative phosphorylation (pyruvate, fatty acids, ADP, P_i, etc.) are transported into the matrix through specific transport proteins (in the physiological literature, these are termed "specific transporters" or "exchangers"; in the biochemical literature, they are called "permeases"). O_2 is thought to simply diffuse into the matrix. NADH, generated glycolytically in the cytoplasm, is not transported directly into the matrix, because the inner mitochondrial membrane is impermeable to NAD^+ or to NADH. Instead, a shuttle system is used to transport the reducing equivalents on NADH to the electron transport chain. ATP as a product of metabolism is transported into the cytoplasm to sites of utilization by processes described in more detail in Figure 1–1; CO_2, another end product of metabolism, merely diffuses out of the matrix.

the fuels being combusted, they are metabolized by pathways that all ultimately feed into the Krebs cycle, where the intermediates can be fully metabolized (Figure 6–2). Being at about the same oxidation state as carbohydrates, the ATP yields of amino acids during oxidation are also similar. For example, alanine oxidation yields 15 ATP/alanine, exactly the same value as for pyruvate. Oxidation of glutamate yields 27 ATPs, while the oxidation of proline could in theory be coupled to the synthesis of 30 ATPs (Hochachka and Somero, 1984).

For practical purposes, then, the above relatively small number of metabolic pathways are the main means by which cells can balance energy production with energy utilization rates. Our ability today to summarize so complex a set of processes so easily and so simply is a ringing endorsement of the research achievements of workers in this area over the last half century. While a great deal of detail remains to be filled in, for our purposes here it is the properties of these metabolic processes which are of greater inherent interest. Like all biological systems, these pathways are the outcome of mutation and selection, and they are thus correctly viewed as "designed" systems, analogous to products of engineering design. Although organisms and metabolic pathways are designed by natural selection while machines are designed by engineers, both design systems share the fundamental features of function and purpose; both also involve trial of variants with the selection of those that work best. Our goal in this chapter is to consider the properties of fuels and pathways that are under such selection. What properties are adaptable over the long or the short term? What properties are advantageous and why? What properties determine limits and constraints and why? To set the stage, let us consider ATP itself.

UTILIZING ATP

In terms of both maximum metabolic rates and in percent activation achievable, muscle tissues are exceptional. During transition from rest to work, ATP cycling or turnover (in flux units defined as μmol ATP $gram^{-1}$ min^{-1}) can increase by orders of magnitude, rather than by small percentage changes. Highest rates of ATP turnover for a range of animals vary systematically between large and small animals. Insect flight muscles achieve the highest ATP turnover rates, over 1000 μmol ATP g^{-1} min^{-1} (Table 4–1). Despite such large differences in ATP fluxes, ATP concentrations do not vary much. Fast-twitch muscles usually contain only about 5 to 8 μmol g^{-1} ATP, while slow-twitch oxidative muscle and insect flight muscles usually contain even less (Guppy et al., 1979; Howald et al., 1978). ATP contents in cardiac muscle are similar to those in skeletal muscle. If only endogenous ATP supplies were available, they would be rapidly depleted during high work rates. Instead, ATP concentrations in muscle and heart are refractory to increased flux in man and in a variety of species tested (see Gadian et al., 1981; Sutton et al., 1981; Sahlin et al., 1978; Driedzic and Hochachka, 1976; Balaban et al., 1990; Hogan et al., 1992; Arthur et al., 1992; van den Thillart et al., 1989; van Waarde et al., 1990). The only way high ATPase rates can be sustained with minimal change in ATP level is through balanced ATP resynthesis. The simplest metabolic pathway for achieving this balance involves phosphagen mobilization. Criteria for a phosphagen to be a useful endogenous fuel have been considered (Hochachka, 1985), and these will be briefly reviewed again.

THE NATURE OF EFFECTIVE PHOSPHAGENS

Phosphagen Storage

All currently known phosphagens are substituted phosphoguanidinum compounds. The two best studied systems are phosphocreatine (PCr) and phosphoarginine (PArg) [formed from creatine (Cr) and arginine (Arg) by creatine phosphokinase (CPK) and arginine phosphokinase (APK), respectively]. Creatine is formed from methionine, glycine, and arginine; in man, this occurs in liver and pancreas, while in some mammals, the first step occurs in the kidney and the second in the liver (Walker, 1979). Creatine is accumulated in skeletal muscles, myocardium, and brain, but apparently mechanisms controlling storage amounts are unknown, although feedback repression of synthesis in the liver may limit total availability at these sites (Walker, 1979). By standards of most metabolites, the total pool sizes of PCr + Cr are high (30 $\mu molg^{-1}$ or more in fast-twitch muscles and somewhat lower in slow muscles and heart). On short- and long-term basis, the amount of PCr available for high power output (burst work) seems to be controlled by hypertrophy, not by concentration adjustments, and this adaptation mechanism seems localized to fast-twitch muscles (Holloszy and Booth, 1976). This is not the case for PArg. It too can be stored at high concentrations (up to 50 $\mu molg^{-1}$ in some species) in this case there seems to be a reasonable correlation between burst work capacities and the amount of PArg stored (De Zwaan, 1983).

Utilizing Phosphagen

For a phosphagen to be useful as an ATP source, its ~P must be transferable to ADP at high rates and at correct times. In vertebrate muscle, this capacity is based upon (i) large amounts of the cytosolic isozyme of CPK, (ii) suitable kinetic properties (Hochachka et al., 1983a), and (iii) correct cellular localization. In the skeletal muscles, CPK occurs in the "soluble" fraction as well as in at least three anchored locations: bound to myosin ATPase, to the sarcolemma with preferential access to Na^+K^+ ATPase (Blum et al., 1991) and to the SR with preferential access to Ca^{++} ATPase (Wallimann and Eppenberger, 1990). It is especially noteworthy that these binding sites are MM-isozyme-specific: other CPK isozymes cannot become "anchored" at these sites. Within any vertebrate species, CPK activities tend to be highest in fast-twitch muscle, lower in slow-twitch muscle, myocardium, and brain; between-species comparisons indicate that relative to other ATP-yielding enzymes, highest CPK activities occur in animals capable of high burst speeds; sluggish species have lower activities. A similar relationship appears to hold for PArg kinases (De Zwaan, 1983).

Net ATP formation from phosphagen is further facilitated by the kinetic properties of these enzymes. During high rates of ATP turnover, the direction of net flux is

$$PCr + ADP \rightarrow ATP + Cr$$
$$PArg + ADP \rightarrow ATP + Arg$$

Even if CPK keeps this reaction close to equilibrium at all times, it is instructive that for the human muscle cytosolic isozymes of CPK, the K_d values for PCr are 72 and 32 m*M* for the binary and ternary complexes respectively, while for ADP they are 0.2 and 0.06 m*M* (Jacobs and Kuby, 1980). Thus, under most *in vivo* concentrations, PCr is not saturating and the enzyme can respond sensitively to changes in [PCr]. On the other hand, CPK affinity for ADP is relatively high, making it so competitive for ADP that at least in early phases of work, rising [ADP] drives the CPK reaction to the right (Gadian et al., 1981). It is not known with certainty for later stages of work whether or not CPK functions under ADP saturation, although some current estimates of *in vivo* ADP levels during activated metabolism are well over 100 $nmolg^{-1}$; i.e., in the range of ADP saturation of CPK. In invertebrate muscles during hard work, ADP levels certainly rise high enough to saturate APK fully (Baldwin and Hochachka, 1985), and this seems to be true during exhaustive work in fishes as well (Dobson et al., 1988; Parkhouse et al., 1988). Once ADP-saturated, the catalytic activity of these kinases must be mainly determined by PCr or PArg supplies or by metabolic modulators, including reaction products (England and Baldwin, 1983). Thus, both enzyme content and enzyme kinetic adaptations of CPK and APK favor ~P transfer to ADP. However, because phosphagen supplies are nonsaturating *and* diminish during bursts of work, these rates must decline rapidly with time.

Phosphagens "Buffer" ATP Content

A third requirement of a useful phosphagen is that mobilization should proceed with minimal effect on ATP pools. In this regard, compounds such as PCr are seemingly well suited for the job because the equilibrium constant (K_{eq}) for ATP formation by the CPK reaction is large: At pH 7.0, the K_{eq} for the reaction is about 2×10^9 (Dawson et al., 1978; Gadian et al., 1981). As a result PCr can be almost completely used up with minimal change in ATP concentrations (See Hochachka, 1985; McGilvery, 1983). In this regard, the thermodynamic properties of other phosphagen kinases differ somewhat from CPK (see Ellington, 1989), suggesting that PCr is thermodynamically advantageous (maintains higher [ATP]/[ADP] ratios under resting conditions).

Phosphagen End Products

A problem with all anaerobic pathways for generating ATP is the simultaneous coproduction of other less desirable end products. On first glance this does not seem to be a problem with phosphagens because none of the end products of phosphagen mobilization appear to be deleterious. Arginine formed from PArg may be an exception because it displays relatively nonspecific disrupting effects on enzymes. This led Somero and his students (Bowlus and Somero, 1979) to suggest that an important function of the octopine dehydrogenase reaction, an LDH analogue found in many invertebrates, is to serve as a sink for arginine. Thus, even here, the end product, directly phosphagen-derived, does not seem to be a problem. However, muscle CPK or APK *in vivo* is coupled to ATPases (mainly to myosin ATPase); in a closed system this means that the end products accumulating are creatine plus P_i. The former is not deleterious because it is involved in no additional metabolic sequences; its main metabolic fate is reconversion to the phosphagen form. Similar arguments can be made for arginine, although it can also be metabolized by other pathways (see Bowlus and Somero, 1979). P_i, however, is another matter. It is a highly reactive metabolite involved in numerous enzyme reactions in intermediary metabolism. As in the case of ATP, where unduly high concentrations must be avoided because of unregulated side effects, the need to control the accumulation of P_i may place stringent limits upon how much phosphagen can be stored as an ATP buffer. Training adaptations thus do not typically lead to any profound adjustments in PCr storage amounts, unless this is achieved indirectly by hypertrophy of fast-twitch muscles or by fiber type replacement.

Osmotic or Ionic Effects of Phosphagen Mobilization

Because phosphagens are stored at fairly high concentrations, the CPK or APK reactions might be expected to lead to ionic and charge perturbations. For example, in the resting state, 30 $\mu molg^{-1}$ PCr^{2-} would require similar amounts of a divalent cation. Cr is, of course, uncharged, and its accumulation would appear to leave the cell with an anion gap. This problem is alleviated by CPK coupling to ATPase, since in terms of charge, P_i^{2-} is equivalent to PCr^{2-}; moreover, the former is a better buffer at near-neutral pH values than all known phosphagens. So, the stoichiometry not only avoids serious osmotic or charge imbalances, but simultaneously minimizes perturbing effects of any $[H^+]$ changes. As far as we know, these features do not change with training regimes. Potential perturbing effects of high phosphate concentrations on free Mg^{++} is a possible problem that may arise, but apparently this issue has not been carefully explored at this time.

Ancillary Roles of Phosphagens

The value of phosphagens as fuels would be greatly amplified if they were also usable in aerobic metabolism. For PCr, at least, such aerobic functions have been probed using a variety of techniques, including introduction of null mutations for the gene for muscle type creatine phosphokinase (van Deursen et al., 1993), and three roles for PCr are suggested; namely, (i) in the shuttling of phosphate metabolites between sites of ADP formation and utilization, a process requiring yet another CPK isozyme, localized to the inner mitochondrial membrane with preferential access to the ATP/ADP translocase (Wallimann et al., 1992; Bessman and Geiger, 1981), (ii) in the preferential channelling of ATP to SR-bound Ca^{++} ATPases (Levitskii et al., 1977) and to sarcolemmal Na^+K^+ ATPases (Blum et al., 1991), and (iii) in the facilitated diffusion of ATP (Meyer et al., 1984). The first two depend upon specific CPK isozymes; the third does not. One or all of these roles (or a combination of them) presumably explains why hummingbird muscle, which is overloaded with mitochondria and displays a minimal anaerobic metabolism, nevertheless contains about *2000 units of CPK* (Suarez et al., 1986). Since the highest aerobic ATP turnover rate in these fast-twitch glycolytic-oxidative fibers approaches the highest anaerobic ATP turnover rate in fast-twitch glycolytic muscle of most vertebrates, these CPK results are instructive: they suggest that maximum CPK and actomyosin ATPase catalytic potentials coadapt.

Whereas similar aerobic roles may be played by PArg in molluscan muscles (Storey and Storey, 1983), they seem to be minimized in insects. Insect flight muscle contains unexpectedly low levels of PArg and of the enzyme mobilizing it. A tracheal-based O_2 delivery system, which can achieve a rapid and efficient rest $\rightarrow$ work transition, is probably the reason insect flight muscle does not require as much ATP buffering by PArg. Mammalian and avian O_2 delivery, on the other hand, is based on pumps and pipes (heart and vessels), which is a slower O_2 delivery system, requiring seconds or even minutes to fully match the maximum aerobic energy demands of the organism (see Connett et al., 1985; Hochachka, 1987; Arthur et al., 1992).

Phosphagens: Their Pros and Cons

From this discussion, it is evident that two potentially important advantages can arise from using phosphagens in support of cell work: *high power output and effective buffering of ATP levels,* especially localized levels near active ATPases (Blum et al., 1991). We can assert with confidence that the first is more important than the second, because numerous invertebrates (England and Baldwin, 1983) and at least one teleost (Dobson and Hochachka, 1987) are known to tolerate very large drops in [ATP] during intense muscle work. Therefore, the *maintenance of [ATP] during very high ATP turnover rates is neither universal nor absolutely requisite.* On the other hand, in most vertebrate muscles, the power output obtainable by using phosphagen as fuel usually is higher than obtainable from any other pathway. Not surprisingly, it is the function most perturbed in muscles of transgenic mice genetically engineered to have no MM CPK expression (van Deursen et al., 1993).

Even in this special context, however, phosphagens show some critical limitations, the most obvious being the 30 to 50 $\mu mol g^{-1}$ maximum storage level (placing a modest ceiling on the total amount of work supportable by this fuel). Again, we can make this assertion with confidence, since some muscles store *less,* but none are known that store more than this amount. While this analysis cannot specify the exact basis for limits to phosphagen storage, the most likely reasons seem to be either (i) that phosphagens are strongly anionic (and if levels were too high, they would begin to create uncontrollable side reactions) or (ii) that the amount of phosphagen-derived P_i could lead to significant metabolic perturbations. Whatever the underlying reasons, the modest limit put on storage amount means that phosphagen-based work is only possible for short times (for about 2 to 5 s at maximum mammalian muscle work rates); any further work, therefore, requires the more universal back-up pathways for ATP replenishment that are also used by cells containing little or no phosphagen. These fall into two categories, anaerobic pathways that form ATP in the absence of O_2 and aerobic ones, which require O_2.

ANAEROBIC GLYCOLYSIS

When O_2 is limiting for muscle work, anaerobic glycogenolysis or glycolysis appears to be the main ''back-up'' mechanism for ATP replenishment after phosphagen supplies are depleted. This metabolic pathway is phylogenetically ancient, and its component enzymes are considered to be highly conservative. Their relatively constant properties are thought to be maintained by rigorous natural selection (Boiteux and Hess, 1981). The only sites where major adaptational changes are known are at the terminal step, normally catalyzed by lactate dehydrogenase (LDH), and at the PEP branchpoint. Thus, many invertebrate groups have evolved different terminal dehydrogenases; the energy yields of these modified glycolytic pathways, however, are unchanged from the classical process of fermentation to lactate. At the level of PEP, some animals possess high PEP carboxykinase activities, which allow a large flow of carbon away from mainstream glycolysis and toward succinate or propionate, as anaerobic end products. These branching pathways may or may not be coupled with simultaneous aspartate fermentation, but in all cases, they are energetically more efficient than classical glycolysis, which is presumably why they are active under extreme hypoxic or ischemic conditions, even in mammals (Hochachka, 1980). As with phosphagens, a minimal set of criteria must be met in order for a compound to be a useful anaerobic fuel for muscle work.

Glycogen—an Ideal Fermentable Fuel

The most basic design criterion for a good fermentable fuel is that it be storable at high and adjustable concentrations. This storage criterion is well met by muscle glycogen. In vertebrate fast muscle, glycogen can be stored at over 100 μmol glycosyl unit $gram^{-1}$; in most vertebrates, glycogen is stored at substantially higher levels in fast than in slow muscles. In some invertebrates with greater relative dependence on this pathway, as in bivalve adductor muscle, glycogen content is so high that it constitutes a large fraction of tissue weight (De Zwaan, 1983; Storey and Storey, 1983). To increase efficiency of storage, some muscles rely upon glycogen–membrane associations or upon large intracellular granules (Hochachka, 1980). In many marine invertebrates, the free amino acid pool is expanded, and several of these (aspartate and the branched chain amino acids in particular) can be fermented at low rates to supplement glycolysis (Collicutt and Hochachka, 1977).

ATP Yields of Anaerobic Pathways

A second important design criterion for good fermentable fuels is that their mobilization should amplify the molar yield of ATP. The ATP molar amplification of glycogen → lactate conversion is 3; i.e., 3 mol ATP per mole of glucosyl unit. The complete fermentation of 100 μmol g^{-1} thus generates at least 300 μmol ATP, and the muscle work possible on this fuel is about ten times that supportable by PCr (because the latter is stored at only 30 μmol g^{-1} and maximally supplies only equimolar amounts of ATP). There is no molar amplification in the aspartate → succinate path and only two-fold amplification in the glucose → lactate path. Whereas ATP stoichiometry of a given pathway is fixed by chemistry, the "mix" of pathways used can in theory be adjusted through training or adaptation. That probably is why glycogen is preferentially used as a fuel for anaerobic muscle work over either aspartate or glucose.

Turning on Anaerobic Glycolysis: Hormones and Neurotransmitters

For the purposes of supporting high intensity muscle work, fermentable fuels should be rapidly mobilizable at appropriate times and rates. On balance, we consider this to be one of the most advantageous design features of the anaerobic glycolytic path. As far as we know at this time, the main reason why sequences such as aspartate → propionate or branched chain amino acids → volatile fatty acids are not favored for anaerobic muscle work is because they cannot sustain high enough flux rates; i.e., they cannot supply sufficient power (see De Zwaan, 1983). In contrast, the high power output supportable by the glycogen → lactate pathway is probably the chief reason why it is almost universally used to back up high intensity (initially phosphagen-supported) anaerobic work. The "flare-up" characteristics of this pathway (i) require *exogenous activating signals (hormones or neurotransmitters)* and (ii) arise from *the regulatory properties of glycolytic enzymes and from the occurrence, at virtually every step in the glycogen → lactate conversion, of a tissue-specific isozyme form or forms.*

The major stimuli or signals for activating muscle energy production are hormones and neurotransmitters, usually acting on a receptor that transduces information across the cell membrane to initiate a branching cascade of intracellular events affecting enzymes, channels, and pumps, the cytoskeleton, and other proteins. A well-studied example is regulation of heart muscle, which we shall use to illustrate the principles of this kind of control. When activities of the cardiac sympathetic neurons are raised by exercise, their nerve terminals release norepinephrine, which binds to β receptors on cardiac muscle cells. In the unstimulated state, membrane localized G-proteins are complexed with GDP and free to diffuse in the membrane plane; when these encounter the hormone-receptor complex, GDP exchanges for cytoplasmic GTP, and the G-protein is thought to dissociate into two mobile components—the activated forms of G proteins. When Ga-GTP encounters adenylate cyclase in the membrane, their association activates the latter and thus the synthesis of cAMP. The function of cAMP, the classical intracellular second messenger, is to activate protein kinase A (PKA), by dissociating it and releasing its catalytic activity. In the final step of this cascade, PKA catalyzes the phosphorylation of a series of target proteins, including (i) glycogen phosphorylase (converting inactive β dimer → active α tetramer and so favoring the rapid activation of glycolysis) and (ii) L-type Ca^{++} channel leading to an increase in intracellular Ca^{++} (and so favoring further Ca^{++} activation of glycogen phosphorylase simultaneously with supplying signal for muscle contraction). This kind of G-protein-coupled signal transduction pathway is, in fact, a common if not the most common means for transferring information across cell membranes (Figure 6–4).

Although such control systems are obviously complex and on first glance seemingly inefficient, their design displays some interesting advantages. One obvious such feature is that the combined steps in information transduction allow enormous amplification of signal. A single occupied receptor may turn on many G-protein complexes; amplification here is determined by the life time of the Ga-GTP complex. A single adenylate cyclase makes many cAMP molecules, and here the signal amplification is truly enormous, because it is a function of the catalytic turnover number of the enzyme! And finally, a single PKA enzyme phosphorylates many serine and threonine residues and thus can even further amplify the original signal intensity. As a result, a few occupied receptors can have a major influence on the cell.

Another advantageous design feature is that the initial events at the plasma membrane can be translated into soluble second messengers. This means that hormonal stimulation can target processes distant from the membrane and can affect metabolic enzymes, contractile proteins, Ca^{++} ATPase, L-type Ca^{++} channels, and even gene transcription. Any given protein only requires the patterns of amino acids around serine and threonine residues that is recognized by PKA to be a target for its action, and many proteins clearly have been selected exactly for this feature. This kind of control system thus allows for a larger divergence of action than would otherwise be possible.

A final advantageous design feature of control cascades such as that initiated by the catecholamines is that they can be shut off at various points. Of such potentially redundant or back-up safety mechanisms at least three are well established. First of

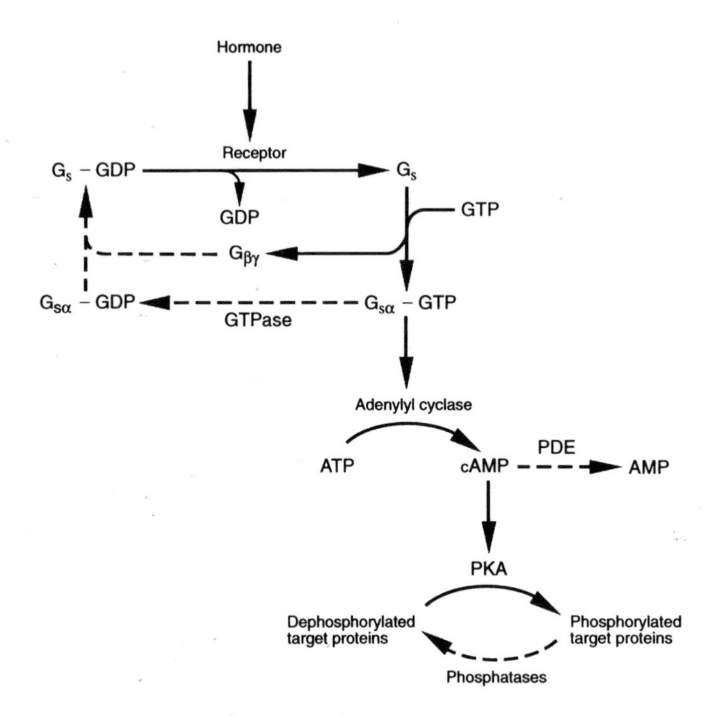

FIGURE 6–4. The action of hormones such as epinephrine is mediated via a membrane-bound receptor and a G-protein-dependent cascade pathway leading to phosphorylation of target proteins (enzymes, channels, transporters, or pumps) and to a change in their catalytic activity state. Sequences involved in transducing the signal from hormone receptor to protein target are shown in solid lines, while dashed lines indicate processes that terminate the signal. Gs, a stimulatory coupling protein, has three subunits identified as subscripts. In other signal transduction cascades, inhibitory G-proteins (Gi) or other G-proteins (Go) may be involved in analogous control systems. Modified from Hille (1992).

all, the G-protein-GTP complex is self-timing and turns itself off by hydrolyzing GTP to GDP. Second, cAMP is cleaved by a specific phosphodiesterase. Third, proteins that have been phosphorylated by PKA action can be dephosphorylated by protein phosphatases. Additional turn-off mechanisms include inhibitory action of G-protein on the cyclase and desensitization of the receptor, a process arising from prolonged exposure to agonist or hormone and that may involve phosphorylation of the receptor per se (Hille, 1992). Thus, a very versatile signal transduction system sets the stage for glycolytic activation in tissues such as muscle and heart.

Turning on Anaerobic Glycolysis: Enzyme and Isozyme Function

With arrival of exogenous signals, the primary controls of flux through the glycolytic path are determined mainly by the catalytic and regulatory properties of isozymes of phosphorylase, phosphofructokinase (PFK), and pyruvate kinase (PK), although other isozymes may also play a role (Hochachka et al., 1983a). Key metabolite activating or deinhibiting signals are supplied by the adenylates (especially ATP and AMP), P_i, and hexose phosphates (especially $F_2,6P_2$).

In addition to the kind of isozymes present, the amounts or concentrations of component enzymes also influence the properties of the glycolytic pathway. In fact, the concentrations of enzyme sites in cytosol fluid, especially of fast-twitch muscles, are surprisingly high and may well exceed the concentrations of many intermediates in the pathway. Recent studies indicate that glycolytic intermediates fall into two categories: (i) precursors and end products, whose concentrations exceed those of enzymes acting upon them and (ii) intermediary metabolites, whose concentrations are substantially less than the enzymes acting upon them. Under these conditions, strong selective advantage would arise from adapting the pathway for preferential enzyme-to-enzyme metabolite transfer of intermediates (i.e., for some channelling, rather than for simple diffusion from enzyme to enzyme in aqueous solution). Experimental evidence for this kind of behavior is available for GAPDH, PGK, LDH and other dehydrogenases (Srivastava and Bernhard, 1986a,b).

An important feature of such enzyme-to-enzyme hand-off systems is that the ratio of reactive substrate and product complexes at any individual enzyme site (K_{eq}^{int}) is near unity; this allows transfer in either reaction direction and in effect smooths out free energy profiles for the overall metabolic pathway. With equal energy partitioning among enzyme-bound intermediates, the unidirectional flux of glycolysis arises because of *the removal of final end products (ATP and lactate) by other enzyme pathways (ATPases in the case of ATP) or by export to a segregated location (efflux out of the cell in the case of lactate or other anaerobic end products).* Srivastava and Bernhard (1986a,b) suggest that selective forces are acting less upon each enzyme component (to increase its k_{cat} or turnover number, so as to maximize flux through the reaction as assayed in aqueous solution) than upon the direct transfer capacities of pathway segments or of the pathway as a whole. Evolutionary pressure seems to favor development of enzyme–enzyme interactions favorable to direct transfer of intermediates. In this view, pathway segments such as pyruvate dehydrogenase (PDH), in which the three enzyme components are covalently linked and intermediates remain bound throughout the catalytic transformation of substrates to products, represent a kind of evolutionary end point in direct metabolite transfer. Srivastava and Bernhard (1986a,b) in fact argue that this is an extremely widespread feature of metabolism, not only a characteristic of the glycolytic pathway. In their view, diffusion necessarily plays a major role in metabolite transfer between functionally distinct compartments but a more minor role within each such compartment.

While stimulation of muscle glycolysis is triggered by a variety of signals (hormones, ions, and metabolites), the mechanisms ensuring that the pathway is correctly phased in with other ATP-generating pathways is frequently under-emphasized. Nonetheless, during anaerobic work, it is widely accepted that phosphagen hydrolysis usually precedes full glycolytic activation; PCr utilization, for example, approaches completion before glycolysis is fully activated. This phase-in behavior arises because, as an ATP synthesizing process, glycolysis is competing with CPK (APK in invertebrate systems) for limiting amounts of ADP. Three enzymes are involved in the competition: CPK (or APK), phosphoglycerate kinase (PGK), and pyruvate kinase (PK). These enzymes are present in the $10^{-4}M$ range (McGilvery, 1975). During early phases of work, both the high activity of CPK and its affinity for ADP ensure that this pathway dominates. On the other hand, the kinetic properties of PGK and PK are well designed for back-up function. In the case of the muscle PK isozyme, for example, the K_d for phosphoenolpyruvate (PEP) is lowered by an order of magnitude from a nonphysiological range to values between 0.01 and 1 mM on addition of ADP. PEP has the same effect on binding of ADP; that is, the binding of one substrate leads to about an order-of-magnitude increase in enzyme affinity for the cosubstrate (Dann and Britton, 1978). This means that as anaerobic work continues and [ADP] rises, PK becomes ever more competitive and facilitates the coupling of myosin ATPase and glycolysis (see Hochachka et al., 1983a).

Because the rate of work supportable by this system is related to the catalytic potentials of the enzyme pathway per se, it is not surprising that athletes trained for short-term high-intensity events have relatively high levels of glycolytic enzymes (and predominantly fast-twitch fibers). It is not yet clear whether the enzyme content per muscle cell changes with training or whether the enzyme profile of any given athlete is a metabolic expression of a unique mix of different fiber types. What is clear is that the enzyme content of this pathway cannot be elevated without limit because of potentially debilitating effects of the end products. It is the need to minimize such problems that defines another criterion of useful fermentable fuels.

The Problem of Anaerobic End Products

Another critical design feature of good fermentable fuels is that they should not generate metabolically deleterious end products. As in phosphagen hydrolysis, the end products of concern are not strictly those formed by glycolysis per se, but from glycolysis-ATPase coupling. In this system, the net reaction and the end products are:

$$\text{glycogen (glucosyl unit)} \rightarrow 2\ \text{lactate}^-$$

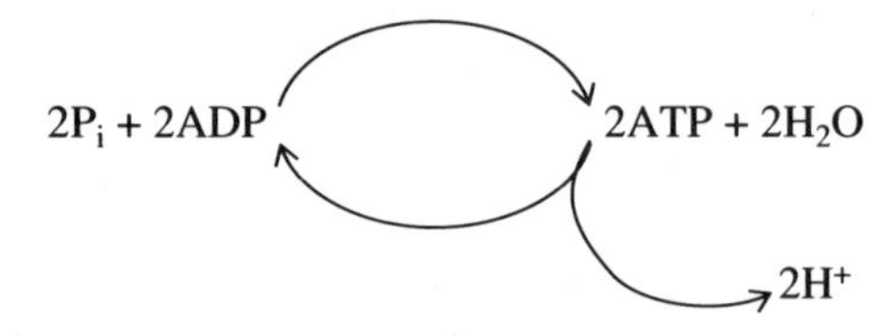

At steady state, two end products accumulate: lactate anions (formed by glycolysis per se) and protons (formed mainly during ATP hydrolysis). Very similar proton stoichiometry prevails for other anaerobic pathways in mollusks or other invertebrates (Hochachka and Mommsen, 1983), the main difference being that n (the number of moles ATP per mole glucosyl unit) varies from 2 (in glucose fermentation) to about 7 in the glycogen → propionate path:

$$\text{glycogen (glucose)} \rightarrow 2\ \text{(anionic end products)}$$

nADP + nPi ⇄ nATP + nH_2O → $2H^+$

Thus, during anaerobic work, two classes of end products accumulate: organic anions (commonly lactate, octopine, alanine, strombine, and tauropine; Gade, 1986) and protons. In principle, both could have deleterious effects. However, compounds such as lactate (which accumulate in mammalian muscle to 20 to 40 μmol g^{-1}) can rise to 200 μmol g^{-1} in animals such as the diving turtle. So, apart from modest osmotic and cation gap problems, these anionic end products present few difficulties for metabolism per se (Hochachka, 1983). H^+ ions, however, are another matter.

In the first place, it is commonly overlooked just how much H^+ is cycled through normoxic metabolism, with no net accumulation or depletion. A simple calculation can show that in a 70-kg man (with a resting metabolic rate of about 700 mmol O_2 h^{-1}) over 100 moles of protons are produced per day. The main sink for these H^+

ions is oxidative metabolism, with a close balance between rates of H^+ production and removal ensuring stable pH. It is well known that the balance of this system breaks down during anaerobiosis, but it is usually also overlooked that *pathways such as glycogen → lactate represent net H^+-consuming reactions at metabolically relevant pH values.* At pH 7.4, for example, this pathway consumes 0.4 H^+ per glucosyl unit, while the pathway glycogen → propionate at this pH consumes an order of magnitude more H^+ per glucosyl unit (Hochachka and Mommsen, 1983); this means that both H^+ stoichiometry and ATP yield are pathway-specific. In both cases, it is the coupling with ATPase that leads to net H^+ production. The metabolically relevant relationship, therefore, is *between H^+ and ATP, not between H^+ and glucosyl units.* For the (glucose → lactate) ATPase coupling, 1 μmol of ATP is cycled per micromole H^+ produced, while for the (glycogen → lactate) ATPase coupling, 1.5 μmol of ATP are cycled through per mole H^+ produced. For glycogen succinate, 2 μmol ATP are cycled per micromole H^+, whereas in glycogen → propionate, up to 3.5 μmol ATP are cycled per micromole H^+ generated (Hochachka, 1983). The strategy of maximizing the amount of ATP turned over per mole of H^+ formed in muscles working anaerobically *at its limit means a balance between H^+ formed and H^+ consumed during ATP cycling, as in the coupling of muscle work to aerobic metabolism* (Vaghy, 1979).

Interestingly, such a balance could be achieved by alcohol fermentations, but such pathways typically do not occur in animal tissues, for they generate their own set of problems (Hochachka, 1983). Thus, when animal tissues are O_2 limited, they necessarily rely upon metabolic pathways that generate H^+ ions, a process that cannot proceed indefinitely.

Upper Glycolytic Limits in Muscle: A Coadaptation Problem

A final question concerns the limits in upward scaling of enzymes in muscle glycolysis. Is there a limit to how much LDH or other anaerobic enzymes animals can incorporate into their muscle cells? The answer apparently depends upon what the muscle is selected for, speed or work. With respect to the former, one of the fastest of anaerobically powered vertebrate muscles is the sound-producing muscle of toadfish (see Hochachka et al., 1988). With reductions in mitochondrial volume densities, this muscle may well be near an inherent limit of about 1:1 for the volume density ratio of SR to myofilaments. ATP synthesizing machinery appears to consist predominantly of enzymes in anaerobic metabolism plus glycogen granules as endogenous substrate. However, because energy demands of this kind of muscle are not high, enzyme activity levels also are not inordinately high. One enzyme activity, that of CPK, however, is notably high (Walsh et al., 1987), suggesting the intriguing possibility of a CPK-Ca^{++} ATPase preferential access coupling of the type found on the modified muscle of the electric organ of electric fish (Blum et al, 1990; 1991). Be that as it may, it would appear that any further upward scaling of SR or myofilaments would be counterproductive because either adjustment would impact on speed, the parameter presumably being selected.

In contrast, most anaerobic skeletal muscles are designed more for power output than for the speed with which contraction-relaxation cycles can be completed. Not surprisingly, the factors determining glycolytic limits are not the same as above. The main difference is in the amount of SR which, compared with the sound-producing muscles, is drastically reduced. Typical values for SR volume densities are in the 4 to 5% range for fast mammalian muscles, only marginally higher than in slow muscles (Eisenberg, 1983). In contrast to reductions in SR and mitochondrial volume densities, the concentrations of glycolytic enzymes are increased in fast-twitch skeletal muscles. The relative directions and magnitudes of these coadaptations can be

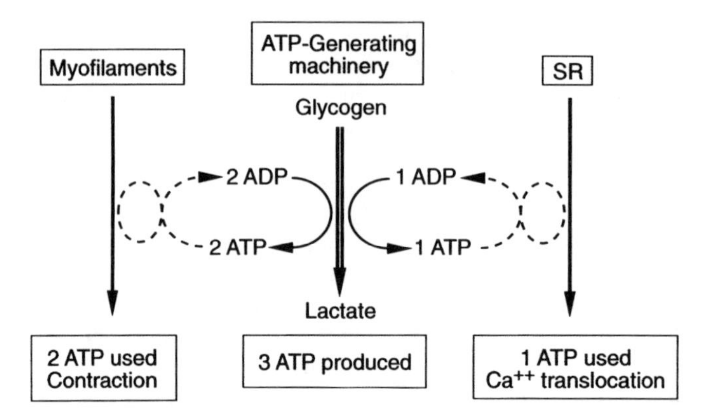

FIGURE 6–5. Upper limit of glycolytically driven ATP turnover in skeletal muscles is determined by the design of the molecular and metabolic systems that utilize and produce ATP.

gleaned by comparative analyses. Thus, in moving from a slow- to a fast-twitch type of mammalian muscle, SR volume densities increase by perhaps two-fold while glycolytic enzyme concentrations may increase by four to five times (Hochachka, 1985). Assuming that glycogen is used as the fuel and that SR function costs 30% of the maximum muscle glycolytic ATP synthesized, then the upper limit for ATP turnover and power output would be attained at glycolytic flux to actomyosin ATPase flux to Ca^{++} ATPase flux ratios of 1:2:1. That is, at steady state, for each glucosyl (from glycogen) converted to lactate, three ATPs would be produced, two of which would be available for actomyosin ATPase, while one would be available for Ca^{++} ATPase (Figure 6–5). Assuming for the moment identical k_{cat} values, the pathway ratios required (expressed as micromoles of substrate converted per gram per minute) would be 1 glycogenolysis unit (producing three ATPs) to 2 actomyosin ATPase units to 1 Ca^{++} ATPase unit, and there would be little or no room left for any further upward adjustment of these three functional units of muscle. Can this limit be attained?

The answer apparently is negative and indeed implies that this limit is undesirable. In rabbit psoas, a predominantly fast-type muscle, myosin concentrations are about 100 μM, actin is at about 300 μM (Yates and Greaser, 1983), and these form some 85% of the volume density of the muscle cell. The k_{cat} for actomyosin is about 10 µmol ATP hydrolyzed (µmol actomyosin ATPase)$^{-1}$ s^{-1}, and thus a maximum value might be about 50 units of ATPase (50 µmol ATP g^{-1} min^{-1}). *In vivo* we know that these rates must be at least 8 to 10 times higher, so a value of 500 units of ATPase is not unreasonable for fast-twitch (PCr powered) muscle fibers. If glycolytic enzymes had similar catalytic capacities, their concentrations would only have to be about half that of myosin (or about 50 μM) to be able to pace maximum muscle work (assuming, as above, that two of each three ATPs are available for actomyosin ATPase). However, the k_{cat} values for most glycolytic enzymes are substantially higher than that for actomyosin ATPase, so the actual concentrations of glycolytic enzymes in theory could be even lower, in some cases well below 1 μM. Paradoxically, however, the concentrations of glycolytic enzymes in fast mammalian muscles are typically 10 to 100 times higher than these minimally expected values (Srivastava and Bernhard, 1986a,b). If the muscle is made faster or the animal is made bigger, even more glycolytic enzymes are packed in.

To confound the problem, these same muscles seem to contain less glycogen than they could, in theory, hold. Generally, vertebrate skeletal muscles contain on the order of 100 μmol glycogen (glucosyl) $gram^{-1}$. In contrast, it is possible to pack in glycogen to well over 1 *M* levels since this level is commonly observed in liver and other tissues of hypoxia-tolerant animals (Hochachka and Somero, 1984). Thus, we are faced with an interesting paradox: fast muscles seemingly contain more glycolytic enzymes than are needed while they maintain glycogen far below possible levels. We feel that this paradox can be resolved in terms of two key features of glycolysis, the accelerator and brake functions, that seem to be under selective pressure in muscle tissue. The accelerator feature is expressed in efficient allosteric regulatory mechanisms and in high activities of glycolytic enzymes, far higher than found in any other tissues of the vertebrate body. From what we know of the kinetic and regulatory properties of muscle glycolytic isozymes (Hochachka and Somero, 1984), these high activities allow (i) rapid flare-up of the pathway and (ii) high flux rates, even though most intermediates of the pathway occur at low concentrations, well below apparent K_m values. Betts and Srivastava (1991) argue that the accelerator function is probably singly the most important selective advantage of high glycolytic enzyme concentration studies. Assuming a $K_{eq} = 1$ at all steps in the multienzyme pathway, their studies show that high enzyme concentrations allow for very rapid on-switching of glycolysis (1000-fold increase in flux in 1 s with modest substrate concentration changes and enzyme concentrations of 25 μ*M*!). Since neither the appropriate concentration changes in pathway intermediates nor uniform K_{eq} values of 1 are likely *in vivo,* there remains a large gap between the *in vivo* situation in muscle and their model. Nevertheless, it does drive home the importance of enzyme concentration in the accelerator function of glycolysis.

Flare-up or accelerator function is not the only property of glycolysis that is regulated. Because glycogenolysis in activated muscle may well operate as an essentially closed system potentially capable of generating harmful end products, it also is necessary to maintain the activated system under close control. While numerous regulatory brake mechanisms, e.g., allosteric enzymes, and defense mechanisms (such as high buffering capacities) have evolved, the ultimate brake feature in many species, especially lower vertebrates (Dobson and Hochachka, 1987) may well be the limited amount of stored glycogen; when this is used up, no matter what else prevails, the pathway must grind to a halt, and it does so before end-product accumulation leads to (i) undesirable acidification, (ii) osmotic perturbation, or (iii) uncontrollable side reactions. In tissues such as the liver, which contain enormously high amounts of storage glycogen, the end product of glycogen mobilization is glucose, not lactate; there is no net generation of protons, and enzyme activities are low. Since the accelerator of muscle glycolysis is absent, the brake also can be slackened.

This analysis suggests that several properties [stored glycogen, catalytic capacities of muscle-specific glycolytic isozymes, actomyosin ATPase capacities (or volume densities of myofilaments), and muscle buffering power] all coadapt; upward or downward adjustments in any one of these parameters requires proportionate adjustment in them all (Hochachka, 1985). That is why we suggest that the upper limit for anaerobic work capacity in muscle must represent a kind of optimization or balancing process that maximizes the amounts of actomyosin ATPase and glycolytic enzymes in appropriate proportions while minimizing the risk by maintaining the right (modest) amount of glycogen and the right (large) amount of intracellular H^+ buffering capacity. At a theoretical upper limit, any further upward adjustment of glycolytic capacity may well require compromise in one of the other coadapting properties, which would be counterproductive. Even the best-adapted glycolytic systems, however, still are critically time-limited, so anaerobic glycolysis must in turn be backed up by aerobic, ATP-replenishing pathways.

OXIDATIVE METABOLISM

The transition from rest to maximum sustained work in mammals requires the ATP turnover rates of the active muscles to increase by some 10 to 100-fold. Requisite ATP is formed by the complete oxidation of carbohydrates, fats, and proteins or amino acids. In contrast to other components of this effector system (see above), most of the enzymes of β-oxidation, the Krebs cycle, and the electron transfer system do not appear to occur in isozymic form; differences between fast- and slow-type muscles are based largely on low or high enzyme content, not on enzyme isoforms. Perhaps in part for this reason the dependence upon enzyme-enzyme preferential transfer mechanisms appears even greater than in anaerobic metabolism. The most obvious reaction sequence illustrating this situation is the electron transfer system (ETS), the enzymes and cofactors of which have long been considered to operate as functional complexes. When NADH is oxidized to NAD^+, two electrons and one H^+ are released. Protons, of course, are soluble in aqueous solution as hydronium ions $(H_3O)^+$. Free electrons, in contrast, do not exist in aqueous solution: to reduce O_2, electrons are passed from NADH or from $FADH_2$ to O_2 along a chain of electron carriers, all of which are components of the inner mitochondrial membrane. At least 12 and probably more electron carriers are grouped into four multiprotein, intramembranous or transmembranous particles or complexes. Although the *in vivo* structural order of these complexes is not known with certainty, the *in vivo* functional order can be determined with certainty with a variety of approaches. For example, *the order of the electron transfer components in the ETS is the order in which they become oxidized after the addition of O_2 to anoxic mitochondrial preparations.* In addition to serving as electron carriers, three ETS complexes also serve as proton pumps, in essence setting up H^+ concentration gradients, which are thought to supply the energy for ATP synthesis (the chemiosmotic theory). How these two primary functions (of electron transfer and of proton pumping) are integrated is not yet well understood; current concepts are available in numerous reviews and biochemistry textbooks.

The importance of structural organization to function is also widely appreciated in studies of the β-oxidation pathway. The interested reader is referred to recent reviews or books on the topic (see Bremer and Osmundsen, 1984). From our perspective, the important point is that β-oxidation is designed to receive long chain fatty acylCoA substrates and release acetylCoA; in the case of palmitylCoA, for example, *none of the 54 intermediates between starting substrate and pathway end products are found in vivo,* which is a classic example of optimized metabolite channelling (Srivastava and Bernhard, 1986b).

Unlike β-oxidation and the ETS, the Krebs cycle reactions are often considered as operating in solution in the mitochondrial matrix. Recently, however, this position has become less clear and certainly controversial. For example, mildly disrupted mitochondria can be prepared, which show exposed if readily sedimentable Krebs cycle enzymes. When either fumarate oxidation or the MDH-CS (malate dehydrogenase-citrate synthase)-coupled enzyme system are analyzed, relative kinetic advantages are observed over fully solubilized enzyme preparations. From these data Srere and his coworkers (Robinson et al., 1987) propose that the Krebs cycle *in situ exists as a sequential complex of enzymes, a metabolon, favoring channelling of metabolites along the metabolic pathway and electrons directly to the ETS* (as in the case of coupling between MDH and NADH ubiquinone oxidoreductase, Complex I in the ETS). Interestingly, the kinetic advantages observed for these coupled reactions are more sensitive to disruption than are the binding interactions with the overall particle; i.e., enzyme-enzyme interactions favoring metabolite hand-offs along the

Krebs cycle are more fragile than are enzyme–inner mitochondrial membrane interactions (Robinson et al., 1987). This is one reason why experimental support for the metabolon concept has been difficult to obtain, and only now is the concept receiving broader attention (not all favorable, we hasten to add!).

An additional, functionally important design feature of oxidative metabolism is that the complete combustion of most substrates and the coupling to ATPases take place in at least two separate compartments, the cytosol and the mitochondria. This introduces problems of exchange between these compartments of (i) carbon metabolites per se, (ii) reducing equivalents, and (iii) adenylates per se—all of which may influence flux rates. [The interested reader is referred to Holloszy and Coyle (1984) and Tager et al. (1983).] Because no single fuel feeds oxidative metabolism and because aerobically working tissues remain perfused, relatively open systems, the criteria for useful substrates depend upon whether the fuel is stored internally (endogenous) or in other depots (exogenous).

Nature of Endogenous Aerobic Fuels

As with anaerobic fuels, the most basic design feature of a "good" endogenous fuel for aerobic metabolism is that it be storable in high and adjustable amounts. The main endogenous fuels in mammalian muscles are glycogen and triglycerides. Both pools are expandable during short-term training or in long-term adaptation (Holloszy and Booth, 1976). Similar fuel preferences are observed in insects, such as locusts and moths (Storey, 1985). In salmon and squid, expendable proteins are dominant energy sources during migration, although there is uncertainty as to pathways of mobilization of the resultant amino acids (Mommsen et al., 1980). The amino acid pool is also utilizable for aerobic metabolism in many marine invertebrates (Storey and Storey, 1983) as well as insects (Storey, 1985).

ATP Yields of Aerobic Pathways

A second, equally important criterion of a good aerobic fuel is that its aerobic metabolism should markedly amplify molar ATP yields. Probably the most striking difference between phosphagens and glycogen in anaerobic ATP generation and fuels for aerobic metabolism *is the immense increase in yield of ATP/mol of the starting substrate* achievable by the complete oxidation of fuels. For triglyceride (calculated for glycerol +3 palmitate), for free fatty acids (calculated for palmitate), for glycogen, for glucose, for proline, and for lactate, the respective molar yields of ATP are 403, 129, 38, 36, 21, and 18, assuming complete oxidation; these values are 18 to 400 times higher than the molar yield of ATP from phosphagen hydrolysis and 6 to 130 times the yield from anaerobic glycogenolysis. The advantages of specializing in different kinds of fuels for different kinds of exercise work are self-evident. Interestingly, such specialization in vertebrates is fiber-type specific. Generally, slow oxidative fibers store higher amounts of triglycerides, while fast-twitch fibers are specialized for storing higher amounts of glycogen. The same coadaptation of fuel and fiber type is evident in interspecies comparisons.

End Products of Aerobic Metabolism

As in anaerobic metabolism, good fuels for aerobic metabolism should not generate undesirable end products. It has been emphasized by Atkinson and Camien (1982) that the main end products of triglyceride and glycogen oxidation are molecular CO_2 and H_2O, while the oxidations of proteins, amino acids, and carboxylates yield CO_2, H_2O and HCO_3^-, as well as NH_4^+. Fortunately, none of these are very deleterious as metabolic end products. Even if NH_4^+ and HCO_3^- can perturb metabolism, such problems can be minimized by their incorporation into urea and by urea excretion. CO_2 can be removed at the lungs or gills, while H_2O is essentially harmless. However, CO_2 as an end product of metabolism cannot be considered in isolation because it is brought almost instantly into equilibrium with H^+ and HCO_3^- by carbonic anhydrase (CA) catalyzed hydration, a reaction that can be written as:

$$CO_2 + H_2O \rightleftharpoons H^+ + HCO_3^-$$

Carbonic anhydrases, which catalyze this reaction, occur in several distinct isozymic forms (Gros and Dodgson, 1988). All skeletal and cardiac muscles possess a membrane-bound CA; fast-twitch (or white) muscle fibers additionally express cytosolic isoform CA II, whereas slow fibers display a kinetically less efficient CA III cytosolic isoform. Carbonic anhydrases are thought to serve at least four different functions in skeletal muscle management of CO_2: (i) facilitate CO_2 diffusion, (ii) extracellular catalysis of CO_2 transport, (iii) intra- and extracellular buffering, and (iv) supplying H^+ or HCO_3^- for SR during Ca^{++} transients.

Facilitated diffusion results from the simultaneous diffusion of CO_2 plus H^+ and HCO_3^-; in fast-twitch or white muscle fibers, the facilitation factor (relative speed-up of CO_2 diffusion) is about 2 and is catalyzed by CA II. In slow-twitch or red muscles, as well as in heart, CO_2 efflux is speeded up by about three-to-four-fold. In the red muscle, the facilitation is due mainly to CA III. In heart muscle, on the other hand, no cytosolic CA isoforms are expressed, yet CO_2 efflux is notably speeded up, showing clearly that membrane-bound CAs are fully capable of facilitating CO_2 diffusion (Gros and Dodgson, 1988).

A secondary function of sarcolemmal-bound CA is to assure that the CO_2-HCO_3^- system is in equilibrium during blood passage through the capillary. Current understanding of how the system works is reviewed by Geers et al. (1991). CO_2 diffuses from the mitochondrial sites of formation by free and facilitated diffusion in the sarcoplasm, then effluxes from the cell by CO_2 permeation of the sarcolemma (which is largely impermeable to HCO_3^-). As it exits the cell, CO_2 is hydrated and held in equilibrium with H^+ and HCO_3^-, a function that requires the active sites of membrane-bound CA to be oriented to the extracellular fluid (ECF).

A third function for muscle CAs relates to H^+ buffering. It is well known that the CO_2-bicarbonate system makes a major contribution to intracellular buffering. This process, which takes up excess protons or releases them according to pH changes, is important both within muscle cells and in the immediate extracellular compartment. It is interesting that white muscles, with a high capacity for lactate and H^+ production, express the catalytically most efficient form of CA (isoform II) and express markedly *higher* levels of sarcolemmal-bound CA than in slow muscle. Since CO_2-HCO_3^- is almost the only major buffer system in the interstitial space, an important role for the high membrane-bound CA in white muscles is preventing drastic acidification of the interstitium during accelerated glycolysis.

Finally, Geers et al., (1991) proposes that CA also may play a fourth and critical role in muscle by providing *a rapid source of, or sink for,* H^+ during Ca^{++} cycling in muscle contraction and relaxation. This hypothesis begins with the observation that H^+ is exchanged for Ca^{++} during Ca^{++} release and uptake across the SR membrane. CO_2-HCO_3^- would be an ideal H^+ donor or acceptor system within the SR due to the literally unlimited availability of metabolic CO_2. However, since Ca^{++} transients (say 100 ms) are some five-fold faster than the uncatalyzed hydration-dehydration reaction, this function requires CA. CA activity within the SR is estimated at about 1000 μmol g^{-1} min^{-1}, easily high enough for CA function to keep pace with Ca^{++} kinetics.

Two other end products of aerobic metabolism that are of concern are alanine and IMP. The former may accumulate during augmentation of Krebs cycle intermediates, but the levels are normally too modest to present any serious problems (see Hochachka and Somero, 1984). Also, under working conditions in mammalian muscle, AMP may be deaminated to IMP, leading to adenylate depletion via the sequence:

$$ATP + H_2O \rightarrow ADP + P_i$$
$$2ADP \rightarrow ATP + AMP$$
$$AMP \rightarrow IMP + NH_3$$
$$NH_3 + H^+ \rightarrow NH_4^+$$

It is instructive that not all animals have this system. In many mollusks and crustaceans, AMP deaminase is absent and [ATP] may drop drastically (below 1 μmol g^{-1}) with a concomitant rise in [ADP] and [AMP] *but with no net change in the adenylate pool* (England and Baldwin, 1983; Baldwin and Hochachka, 1985). In exhausted trout, muscle [ATP] is similarly allowed to drop (to as low as 0.5 μmolg^{-1}), but in this case, because AMP deaminase is present, there is also a large reduction in the adenylate pool (Dobson and Hochachka, 1987). Although such reductions in muscle ATP content rarely if ever occur in human muscle, the above results imply that the work obtainable from ATP hydrolysis remains adequate at quite widely varying adenylate levels, with or without AMP deaminase function, with or without "defending" [ATP] at a threshold of 2 to 3 μmolg^{-1}, as occurs in working muscles in man, with or without ADP and AMP accumulation (Dawson et al., 1978).

The notable interspecies differences raise serious questions on the functions of AMP deaminase in species that possess it. The most likely answer is that by supplying IMP for adenylsuccinate synthetase, AMP deaminase sets the stage for fumarate formation from aspartate and, thus, for augmenting Krebs cycle intermediates when they are needed. This would explain the occurrence of this enzyme in vertebrate muscles, as well as its absence in invertebrates, where proline and glutamate are abundant and serve anapleurotic roles (De Zwaan, 1983; Storey and Storey, 1983). It also explains why IMP is a potentially adaptive end product of aerobic metabolism.

Nature of Exogenous Fuels of Aerobic Muscle Metabolism

Storage Sites of Exogenous Fuels

Whereas most of the ATP turnover for high intensity and relatively short-term work (minutes to hours) is primarily sustained by endogenous fuels, long-term exercise work (many hours to days or weeks) necessarily requires exogenous fuels which, by definition, should be storable at high and adjustable amounts at sites other than working muscle.

This criterion in vertebrates is met for all major classes of storage carbon substrates. For glucose, this criterion is met by liver glycogen depots, for FFA, by adipose fat. When lactate and amino acids are used as fuels, they are mobilized mainly from glycogen and protein stores in nonworking muscles (see Hochachka and Somero, 1984; Hochachka and Guppy, 1987; Brooks, 1986; Hochachka, 1986; and discussion on lactate below). In ruminants, the situation is complicated by the rumen of the digestive system, in which short-chain volative fatty acids, especially 2-, 4-, 5-, and 6-carbon chains, are the dominant carbon and energy sources entering the circulation. The metabolic complications occur mainly in the liver, where these fuels are partitioned between glycogenesis, lipogenesis, and oxidation (Hochachka, 1973).

In contrast to carbon substrates, O_2 is an exogenous substrate that cannot be stored in any appreciable amounts, and for this reason mechanisms for assuring high rates of transport must be provided in active animals (see below).

Flux Rates Sustainable by Exogenous Fuels

Because aerobically working muscles are perfused open systems, a critical feature of any exogenous substrate is that it must be capable of sustaining adequate flux rates. In units defined as $\mu mol\ g^{-1}\ min^{-1}$, *the flux rates required depend upon the molar yield of ATP*. Thus, an ATP turnover rate in human muscle of 20 $\mu mol\ g^{-1}\ min^{-1}$ could be supported by a palmitate flux of 0.15 units, a glucose flux of about 0.6 units, a lactate flux of about 1.2 units, and an O_2 flux of about 3.3 units. The more energetically efficient a fuel is (in terms of ATP/substrate burned) the less rapidly it need be fluxed from depot site to working muscle. For this reason, enzymes mobilizing efficient fuels (e.g., acylCoA synthetases) can occur at lower activities per gram tissue than those mobilizing relatively inefficient fuels or those operating in anaerobic metabolism.

Of these fuels, the highest flux is required by O_2, which is another important reason why its transfer from lungs or gills to tissue mitochondria is facilitated by hemoglobins (or their analogues) and by muscle myoglobins. In fact, very recent studies suggest that a previously unrecognized role of muscle myoglobin may involve a preferential access of O_2 from myoglobin to cytochrome oxidase (Wittenberg and Wittenberg, 1987, 1990; White and Wittenberg, 1993). This process is formally similar to the glycolytic hand-off mechanisms considered above, and on its own could speed O_2 fluxes.

Despite such transfer-facilitating mechanisms, many physiologists and metabolic biochemists conclude that, in vertebrates generally, O_2 availability becomes limiting at maximum aerobic metabolic rates (for example, during maximum sustainable exercise). Some believe it is limiting because the cardiac pump is limiting. This argument emphasizes that higher rates of O_2-based muscle work are attainable when only a small muscle mass is involved than when the whole organism is exercising. In species where this is so, man for example, maximum whole-organism work would therefore not be limited by O_2 uptake capacities at the muscle per se but by the heart's capacity to circulate O_2-loaded blood (Saltin, 1985). In some species, dog for example, the pump capacity appears to be relatively expanded, and the available evidence suggests that O_2 uptake capacities at the working muscles contribute to setting the highest whole-organism sustainable work rate. Recent analyses indicate that O_2 limitation (either delivery- or diffusional-based) is probably the rule during maximum whole-organism work in organisms with lungs or gills but may not be so in insects. In the latter, O_2 delivery by tracheoles is so efficient that the *in vivo* cytochrome oxidase turnover numbers are reasonably close to the theoretical maximum rates, in contrast to vertebrates, where cytochrome oxidase *in vivo* works at only a modest fraction of its theoretical maximum (see Hochachka, 1987; Suarez and Moyes, 1992). In any given mitochondrial-based muscle system, however, the rules of the game—involving trade-offs between the up-regulation of the two main ATP demand pathways and mitochondrial ATP supply pathways (Figure 6–6)—are the same as for endogenous fuels. If frequency is being selected for, the SR Ca^{++} ATPase energy demand system is greatly expanded at the cost of the myofilaments; if strength is selected for, then the myosin ATPase (myofilament volume densities) cannot be compromised. In both systems, mitochondrial volume densities can be up-regulated to points at which they begin to negatively affect either myosin filament volume or SR volume or both. Within this general framework, O_2 may play a special role in the

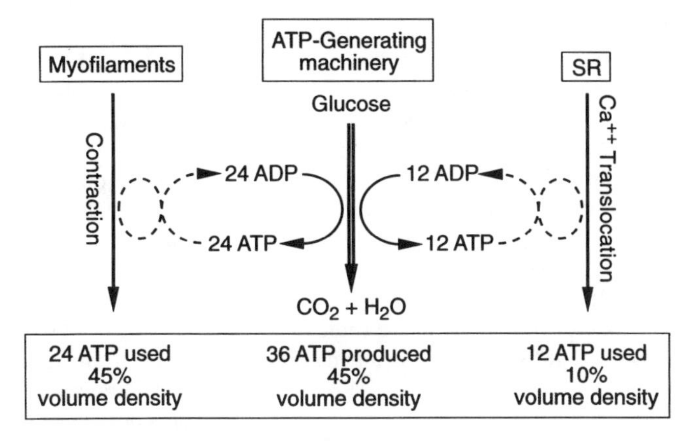

FIGURE 6–6. Upper limits of aerobic ATP turnover systems in skeletal muscles are determined by the competing requirements of the ATP-utilizing and ATP-synthesizing machinery.

integration of ATP demand and ATP supply pathways, a topic that is further explored in Chapter 7.

Regulating Fluxes of Exogenous Fuels

A final and critical design feature of good exogenous fuels is that their fluxes from storage tissues to working tissues should be adjustable according to needs. In general, uptake at the working muscle varies with delivery rate (perfusion) and with plasma [substrate]. This feature requires provision for plasma substrate availability varying with working needs. Plasma glucose often does not satisfy this provision at low work rates, but may do so at high rates (Jones et al., 1980). Plasma FFAs, on the other hand, increase in availability during modest exercise, making them a quantitatively important fuel (Maughan et al., 1978). In man, their contribution decreases during heavy exercise (Jones et al., 1980), while in llamas, their contribution increases with prolonged work (Hochachka et al., 1987). Similarly, insects such as locusts begin long-term flight on carbohydrate, but as work continues, they rely more and more on fat (Storey, 1985).

Most earlier analyses of fat metabolism have overlooked the potential role of peroxisomes in the overall mobilization process. According to current concepts, one function of peroxisomes is to partially degrade long-chain fatty acids (C-16 and longer) to medium- and short-chain fatty acids. The latter are then either transferred to working tissues or are fully oxidized by mitochondrial metabolism *in situ*. The advantage of this arrangement seems to arise from (i) higher water solubility of shorter-chain fatty acids and (ii) higher rates of mitochondrial oxidation than sustainably by long-chain analogues. This area is in a rapid stage of development; suffice it to mention here that this function of peroxisomes seems to be a feature of many tissues. In the rat, the relative abundances are heart > liver > diaphragm > kidney > skeletal muscle (Veerkamp and Moerkerk, 1986).

A final oxidative fuel to be considered is lactate per se. Although lactate has been emphasized as a muscle fuel from time to time (Stainsby and Welch, 1966; Issekutz et al., 1976), all but a handful of workers have overlooked the magnitude of its contribution. In most recent analyses (see Hochachka et al., 1987; Brooks, 1986; Hochachka, 1986), it is supposed that lactate is formed in sites such as fast-twitch glycolytic (FG) fibers at about the same rates as it is utilized in more oxidative-type fibers, and that, at the upper limit, plasma lactate replacement rates equal oxidation rates in aerobically working muscles. In rats, as much as 50% of the carbon flux to CO_2 (during maximum sustained muscle work or at high lactate levels) accords with such a lactate ''shuttling'' model, while in dogs and tammar wallabies, the maximum value is 20 to 30% (Hochachka et al., 1985).

These lactate flux rates are only high by the standards of aerobic pathways; anaerobic glycolytic fluxes can be two orders of magnitude higher. Because of its energetic inefficiency, however, anaerobic glycolysis hardly affects total ATP turnover. In wallabies, for example, glycogenolysis in FG fibers (adequate to account for the maximum flux of lactate to aerobically working muscles) would contribute only 1 to 2% to the total ATP cycling rate. These rates of glycolysis [which may be anaerobic or aerobic (Brooks, 1986)] are similar during aerobic exercise to those in other animals (Taylor et al., 1981; Brooks et al., 1984) and lead to an interesting conclusion, namely that during sustained exercise the *function of glycogenolysis is less to make ATP at lactate production sites than it is to mobilize glycogen (via lactate) as a fuel for aerobically working muscles* (Hochachka et al., 1985, 1987; Weber et al., 1987). If, in a 70-kg man with 30 kg of muscle, 10 kg were used for steady work rates, while 10 kg were used intermittently, it is easy to show that enough glycogen is available (100 μmol g^{-1}) to sustain observed lactate turnover rates of 500 μmol kg^{-1} min^{-1} for about 3 h, while with training, the work time could increase (due to higher amounts of stored glycogen). Hence, the supplies of precursor are theoretically adequate to meet a significant energy need, which may be why during endurance exercise in man, lactate is a potentially useful fuel (Brooks, 1986), while the use of plasma FFA seems to slow down with time (Jones et al., 1980). However, this sequential change in fuel use is not seen in all animals. In llamas, lactate flux can contribute significantly to short-term aerobic work, but after 30 min even at modest exercise [less than 50% of $\dot{V}O_{2(max)}$], lactate availability drops to less than at rest and so makes a minimal contribution to muscle energy demands. At this time, [FFA] in the plasma and FFA fluxes both rise and largely account for muscle ATP turnover rates (Hochachka et al., 1987).

The ''shuttling'' role of lactate between sites of formation and site of utilization also is minimized or totally lacking in fishes and insects. In fishes, lactate is formed in large amounts in white muscle during swimming exercise. However, unlike the situation in mammals, lactate seems to be preferentially retained in white muscle, and lactate clearance takes many hours of recovery when perfusion of muscle is well above normal (see Milligan and Wood, 1986). While this pattern has been known for over three decades, the molecular basis for it is not fully appreciated even today. One popular theory is that the glycogen → lactate pathway is arranged to discharge during burst work, while during recovery, most of the lactate is merely reconverted to glycogen *in situ,* a kind of discharging-recharging ''spring coil'' view of glycogenolysis. Current evidence and literature is reviewed by Milligan and Wood (1986), Moyes et al., (1992), and Schulte et al., (1992).

In flight muscles of insects such as the locust, LDH for practical purposes is deleted (Sacktor, 1976; Storey, 1985), so that muscle function here is obligately aerobic, and there is no reliance on the lactate ''shuttling'' role found in mammals. Muscles in very small mammals (shrews) and very small birds (hummingbirds) likewise display relatively reduced activities of LDH, which is apparently retained as a mechanism for buffering cytosolic redox particularly during rest → work transitions in a manner analogous to CPK buffering of the ATP pool under similar conditions (Hochachka et al., 1988). Presumably, this function can be abandoned in insect flight muscle because the tracheole-based O_2 delivery can match the square-wave muscle work transition from rest to flight, a situation that cannot be achieved in organisms that circulate O_2 from lungs or gills to working tissues (Hochachka, 1987).

Coordinating Aerobic and Glycolytic Pathways

The final design criterion of oxidative metabolism that needs evaluation is the means for integrating it appropriately with anaerobic metabolism. In this regard, it is widely observed *in vivo* that as exercise intensity increases, $\dot{V}O_2$ rises to a maximum level, at which time further muscle work depends upon anaerobic glycolysis. How is this integration of aerobic and glycolytic pathways achieved? Although many regulatory signals may be involved (see Hochachka and Somero, 1984), in mammals, at least during high intensity sustained exercise, O_2 delivery to working muscles may represent an important limitation and thus signal when anaerobic metabolism should be phased in. The argument for this conclusion depends upon the observation that muscle ATP turnover rates can be higher in muscles working in isolation than *in vivo*. For example, in one-legged exercise in man, work rates ranging up to 100 μmol ATP g^{-1} min^{-1} can be sustained by a fully oxidative metabolism; work at this very high rate for the entire muscle mass of a man would require a cardiac output some three-fold higher than possible. Thus, as pointed out above, many exercise biochemists (see Saltin, 1985, for example) argue that *in vivo,* the cardiac pump, and thus O_2 delivery to muscle mitochondria, is a major determinant of control. In this kind of situation, then, O_2 concentration in the muscle would signal when anaerobic glycolysis needs be strongly activated.

These studies are instructive because they predict that during intense work involving only a relatively small muscle mass, O_2 delivery to muscle mitochondria should not be limiting. Experimentally, this prediction in fact seems to be realized. During heavy (but not maximal) work in isolated dog gracilis, for example, recent studies show that autoregulation mechanisms can maintain O_2 delivery to muscle at high enough rates to saturate mitochondrial cytochrome oxidase (Connett et al., 1985; Gayeski and Honig, 1986). In this situation, as in insect flight muscle (Hochachka, 1987), O_2 concentration presumably has minimal control influence (however, see O_2 sensing roles in Chapter 7). If O_2 concentration is not the signal for phasing in glycolysis, what is?

Most workers consider that another important control mechanism revolves around acceptor control of muscle mitochondrial function. According to current thinking (e.g., Chance et al., 1981), mitochondria in nonworking muscle are in state 4; flux rates through the electron transfer system (ETS) are low because of limiting [ADP] or [P_i]. During muscle work, mitochondrial respiration and phosphorylation rates increase (the mitochondria are said to enter state 3) as [ADP] and/or [P_i] increase. While the overall response *in vitro* displays positive site-site cooperating in catalysis (Matsuno-Yagi and Hatefi, 1985), at low substrate levels, respiration increases nearly directly with [ADP] or [P_i] (Jacobus et al., 1982), which is presumably also why [with *in vivo* apparent K_m values for ADP of about 22 to 28 μM (Chance et al., 1986)] work rates seem to vary in a linear fashion with [P_i]/[PCr], also quantified *in vivo* (Chance et al., 1981). Although these interpretations are consistent with studies of isolated mitochondria (Jacobus et al., 1982), the *in vivo* data on skeletal muscle are more controversial, and such simple acceptor-based control also clearly does not apply to cardiac muscle. In the latter, work and metabolic rates can increase by at least five to seven-fold with no measurable change in NMR-visible phosphate compounds. In this case, NADH concentration changes may contribute to regulation of heart metabolic rate transitions (Balaban et al., 1990). The *in vivo* studies implying a role for ADP in metabolic regulation, however, do underline the advantage, with endurance training, of increasing the capacities of the electron transfer system (ETS) and oxidative phosphorylation (Davies et al., 1981); for any given submaximal work rate in trained muscle, *flux through the ETS per gram will be lower, because the amount of ETS per gram is up to two-fold higher.* Muscle mitochondria from endurance-trained individuals, therefore, almost necessarily operate at lower [ADP] or [P_i]; i.e., on the steep parts of the ADP or P_i saturation curves. Second, because *in vivo* operation appears to occur at lower concentrations of acceptors and because the maximum capacity for state 3 flux is elevated, the overall metabolic scope for activity is effectively expanded. Finally, when the state 3 capacity is surpassed in either trained or untrained muscle, [ADP] and [P_i] may rise even further, creating conditions favoring glycolytic competition for both. This would ensure that in all species, despite great variation in speed capacity or size (Seeherman et al., 1981), anaerobic glycolysis is phased in at the appropriate time: when the ATP replenishing capacity of oxidative phosphorylation is being outstripped by cell ATP demands.

7

Integrating ATP Supply and Demand

QUANTIFYING ENERGY COUPLING

In our discussions so far, the importance of maintaining energy demand in balance with energy supply has been reviewed in very qualitative terms. This, in fact, is typical of publications in this area of research, and almost no studies have focused on quantifying the tightness of energy coupling. The main insights into this problem from previous studies (e.g., Balaban, 1990; From et al., 1990) come from metabolite concentrations changes, especially changes in [ATP], [ADP], [AMP], [P_i], [PCr], [creatine] (Cr), [H^+], and [lactate] to gain an estimate of degree of imbalance in ATPase fluxes and ATP synthase fluxes. Mostly, these studies of metabolic regulation focus on ATP-yielding pathways and tend to overlook or ignore regulation of ATP utilization (see Rumsey et al., 1990; Arthur et al., 1992). In contrast, the biologically relevant ''pathways'' for working muscles *in vivo* necessarily include the latter, and the most meaningful end products of the pathway in active muscles are work, force development, or power output, not CO_2, H_2O, lactate, H^+, or creatine. Similarly, the most crucial metabolic rate is ATP turnover rate, not O_2, CO_2 or carbon fuel fluxes. ATP turnover rate, of course, is a composite of ATPase and ATP synthase fluxes, and the regulatory integration of these two reaction pathways determines the tightness of energy demand-energy supply coupling. If energy coupling is very tight, ATPase fluxes are closely similar to ATP synthase fluxes, and [ATP] will not change. If energy coupling is loose, ATPase fluxes exceed ATP synthases, and ATP concentrations clearly must fall. Thus, one measure of coupling efficiency is the change in [ATP]/ATP turnover rate, and we can define the percent imbalance in energy coupling over any given work time period as follows:

$$\%\ \text{imbalance} = \frac{\Delta\,[\text{ATP}] \times 100}{\text{total ATP turnover}}$$

The lower the percent imbalance in ATPase fluxes compared to ATP synthase fluxes, the tighter (or the more efficient) the coupling of the two pathways.

A difficulty with this approach is that even during intense exercise, in mammalian skeletal muscles [ATP] almost never declines below about 50% of rest values; thus, under many circumstances, estimates of [ATP] change are based on differences between two fairly large similar values. For this reason, in muscles that have high AMP deaminase activities, inosine monophosphate (IMP) is sometimes a better measure of the degree to which ATPase fluxes are out of balance with ATP synthase fluxes. This is because in vertebrate fast muscles, IMP is formed from AMP in a reaction catalyzed by AMP deaminase, and its accumulation in human muscles (Graham et al., 1990), as in all vertebrates (Mommsen and Hochachka, 1988), is stoichiometric with ATP depletion. Since IMP in nonworking muscle occurs at very low concentrations, the percent change in concentration during intense exercise is proportionately much greater than the percent change in [ATP]. The stoichiometry between ATP and IMP, however, implies that changes in the concentration of either (as a fraction of overall ATP turnover rate) yield essentially the same information on the percent imbalance in ATPase and ATP synthesis rates over any particular work period.

Because creatine phosphokinase (CPK) is generally assumed to function under near-equilibrium conditions and to "buffer" changes in [ATP], one might consider that CPK-catalyzed decreases in [PCr] could supply an additional means of estimating percent imbalance in energy coupling; that is, percent energy imbalance could be defined as (Δ[ATP] + Δ[PCr]) 100/ATP turnover. While acceptable, this interpretation may be less appropriate than ours since, at the enzymatic level, energy coupling concerns ATP, not related pathway intermediates. In terms of stabilizing energy coupling, the glycolytic and mitochondrial metabolic pathways also clearly serve in "buffering" change in [ATP], although by different mechanisms than in the case of CPK-based ATP "buffers". Furthermore, in fast-twitch muscles, [IMP] supplies the most unequivocal measure of energy coupling, but its concentration changes are not stoichiometric with [PCr] changes, while they clearly mirror [ATP] changes (Schulte et al., 1992). For these reasons, we generally prefer to quantify efficiency of energy coupling by focusing directly upon ATP or IMP, assuming PCr to be an ATP source analogous to other ATP yielding pathways.

ENERGY COUPLING IN ANAEROBICALLY-DRIVEN MUSCLES

To put this problem into context, it is useful to begin by quantifying energy coupling efficiency in fast-twitch glycolytic muscles, which under burst work conditions are powered primarily by anaerobic ATP, producing pathways. Two recent data sets on trout and tuna white muscles, respectively, are particularly complete (Schulte et al., 1992; Arthur et al., 1992b). Thus in trout swimming to complete exhaustion, the change in PCr and lactate concentrations represents an ATP production equivalent to about 91 μmol ATP per gram muscle over the entire work bout. Although the exact contributions of aerobic and anaerobic metabolism to white muscle work under these conditions is not known, most workers in this area assume that the system is largely driven by anaerobic metabolism (see Schulte et al., 1992). We will assume three different situations with aerobic contributions of 80%, 50%, and 20% respectively. In the first instance, this would mean a total ATP production of 455 μmol ATP per gram; in the second, it would mean a total of 182, while in the third, it would mean a total of 109 μmol ATP per gram over the work period. During this time, IMP concentrations increased by about 5 μmol/g. Thus, the percent imbalance in the first case would be 5(100)/455 or about 1.1%; in the second and third instances, it would be 2.7% and 4.6%, respectively. Given what is currently known about white muscle metabolic organization (Moyes et al., 1991), the range 2.7 to 4.6% efficiency of energy coupling for trout white muscles is probably realistic. Similar calculations for tuna white muscle (which is substantially more oxidative that in the case of trout) show that at exercise to fatigue, [PCr] changes are similar, but lactate accumulation reaches 150 μmol/g, indicating about a 1 to 2% efficiency of energy coupling (Arthur et al., 1992b). Thus, for muscles whose power output is 50% or more dependent upon anaerobic sources of ATP, the percent imbalance in ATPase fluxes over ATP synthesis rates is in the range of 2 to 5%, and, as we shall see below, this drops substantially as the muscle is designed to be more and more dependent upon oxidative metabolic pathways.

HUMAN MUSCLE AT MAXIMUM AEROBIC WORK RATES

Perhaps the highest mass-specific O_2 update rates achieved by human skeletal muscles occur during work of small-muscle masses. Presumably because of the very high perfusion rates that are possible, O_2-based metabolic rates equivalent to about 105 μmol ATP turned over per gram skeletal muscle per min can be sustained for periods of about 10 min (Andersen and Saltin, 1985); glycolytic plus PCr contributions are < 0.2% of this, about 167 μmol ATP per gram muscle (Graham et al., 1990), for a total ATP turnover during the exercise protocol of 1217 μmol ATP per gram muscle per 10 min. Over this time period, ATPase fluxes modestly exceed ATP synthase fluxes because the concentrations of ATP when fatigue is reached under these conditions drop by about 2.1 μmol ATP per gram muscle. For the total ATP turnover, the percent imbalance is about 2.1 × 100/1217 or 0.17%, representing coupling efficiency at least several-to-ten-fold higher than observed in more anaerobically-dependent fast-twitch glycolytic muscles. This estimate is close to that for rat muscle during treadmill exercise, where the percent imbalance in fast-twitch oxidative (FOG) muscle (driven almost exclusively by O_2-based metabolism) is estimated at between 0.06 and 0.12% (Hochachka et al., 1991). Although for O_2-based work, the efficiency of energy coupling is adaptable (and can change with training, acute hypoxia, chronic hypoxia acclimation, or long term altitude adaptation (Matheson et al., 1991), in all these cases, the precision of flux regulation in the two directions is impressively high. It is even higher at lower work rates where this kind of quanification of energy coupling is impossible (because [ATP] changes are too small to be accurately measured).

If we phrase the problem in terms confronted by the enzymes involved (i.e., molecular rather than molar terms), during calf muscle exercise in Andean natives, for example, about 3.3×10^{20} molecules of ATP per gram muscle are converted to ADP and P_i, then back again before fatigue. In the process, only 20 to 25% of the ATP pool is lost to IMP (Matheson et al., 1991), and the situation is even better at low work rates! The question arising is how this unexpectedly enormous regulatory precision is achieved.

SETTING THE ATP DEMAND: *V* vs *S* or *S* vs. *V*

To explore the problem, let us begin with first principles. Classically, enzyme kinetics and regulation are analyzed in terms of reaction velocities (v) as functions of substrate concentrations, [s]. For Michaelis-Menten kinetics, enzyme activation usually involves increasing while enzyme inhibiton involves decreasing apparent enzyme-substrate affinities, with or without effects on V_{max} (Figure 7–1a). For allosteric or regulatory enzymes displaying sigmoidal substrate saturation curves, activation and inhibition similarly are largely based on enzyme-substrate affinity changes (Figure 7–1b). These analyses assume that [s] is the independent parameter, while v is the dependent parameter, which may be valid for many tissues, but (most physiologists would maintain) not for working muscle. To physiologists, the ATP demand of a working muscle is considered to be mainly determined (i) by recruitment of motor units and (ii) by rate coding (Hoffer et al., 1987). Once any given motor unit is recruited, energy demand is presumably determined largely (if not solely) by the firing frequency (activation intensity), which at the cellular level effectively means Ca^{++} interaction with troponin *c* and consequent actomyosin ATPase activation. In these terms, activation intensity presumably is a complex function of Ca^{++} transient and Ca^{++} concentration per se (Ruegg, 1986), but activation does not depend

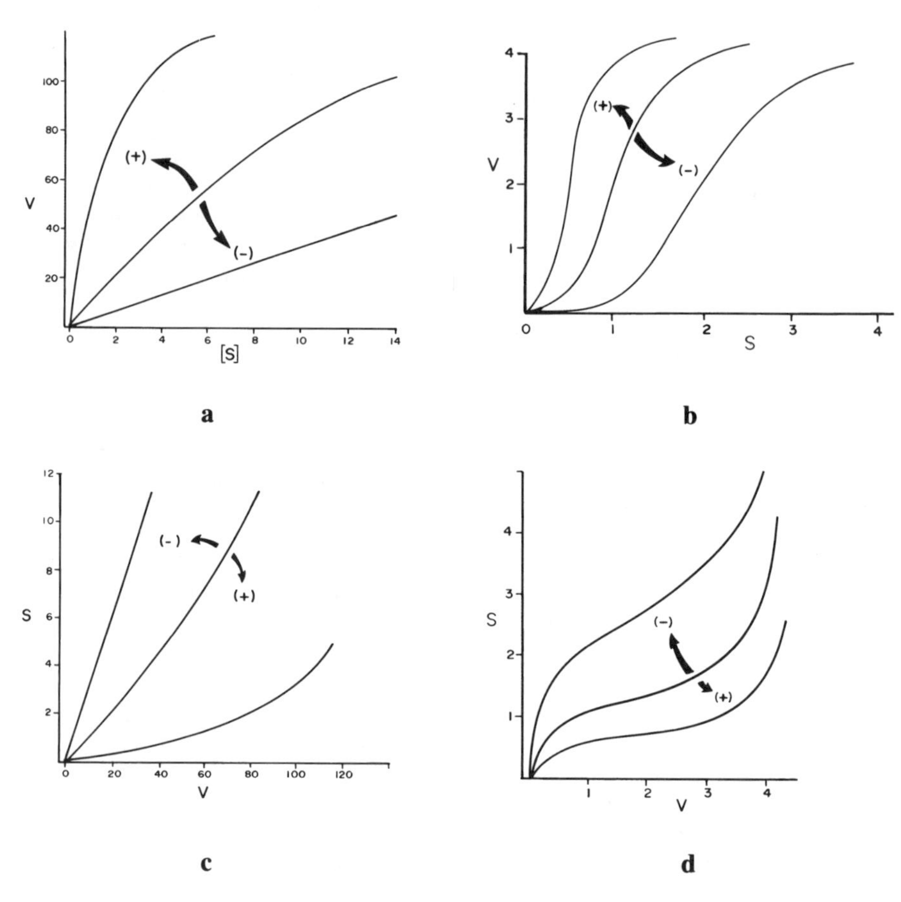

FIGURE 7–1. Plots of [*s*] vs. *v* and of *v* vs. [*s*]. (a) Michaelis-Menten plot with [*s*] as the independent parameter. (b) Sigmoidal saturation curve with [*s*] as the independent variable. (c) Michaelis-Menten plot with *v* as the independent parameter. (d) Sigmoidal saturation with *v* as the independent parameter. Effects of positive modulator (+) and of negative modulators (−) are shown by dark arrows.

on a rise in the global concentration of the substrate, ATP. As stressed above, the affinity of Mg^{++} myosin ATPase for ATP is so high (affinity constant in the n*M* range) that the enzyme is, in effect, saturated (or nearly saturated) with respect to ATP under most physiological conditions (Hackney and Clark, 1985). That, too, is the situation with respect to actin, which can be viewed as a cosubstrate for myosin ATPase, but which is moved rather than chemically converted during enzyme function (actin determining direction of sliding of filaments; myosin determining the rate (Sellers and Kachar, 1990)). In functional terms, this means that in resting muscle myosin, ATPase is *a largely inert enzyme, latent but primed for catalytic action, awaiting only the arrival of an exogenously generated activation signal.* Since the entire pathway of ATP turnover is initiated at this point by an exogenous signaling path potentially extending all the way back into the central nervous system, most biologists would accept that this clearly represents a proactive control mechanism. For any working situation faced by the organism, it represents the means for setting the ATP demand of exercise, albeit tempered by product (diprotonated phosphate) inhibition (Cooke and Pate, 1990; Hogan et al., 1992; Wilkie, 1986). This means that *for any metabolic enzyme involved in ATP regeneration, the required flux is set by the ATP demand and substrate concentrations are adjusted accordingly; i.e., reaction velocity (v) is the independent parameter, while substrate concentration [s] is the dependent variable.*

This kind of representation, where *v* determines [*s*], while unusual in the literature on enzyme kinetics and regulation, nevertheless is instructive (Figures 1c and 7–1d). In the first place, it clearly shows for both regulatory and Michaelis-Menten enzymes that activating conditions are characterized by *lower* substrate concentrations than control or inhibitory conditions. Second, high fluxes may be characterized by lower or almost unchanging substrate concentrations compared to control conditions. Third, and most important, large changes in flux (or reaction velocity) can occur with modest change in [substrate], particularly for regulatory enzymes (Figure 7–1d). Finally, these *s* vs. *v* plots (Figure 7–1c,d) indicate that *in vivo* velocities are likely never to approach V_{max}, for the obvious reason that substrate concentrations begin to rise exponentially and uncontrollably at the approach to saturation.

Characteristics of enzymes exposed when *v* is the independent parameter are recognized from time to time by metabolic biochemists (Atkinson, 1990), but they are unheard of in physiology and are probably counterintuitive for most biologists and physiologists (who would normally expect increasing concentrations of pathway intermediates to cause increasing flux); that is why we will review some standard experimental studies to see which paradigm can best accommodate the data. At the outset, however, we should emphasize that, as shown in the overview in Chapter 6, most current models of metabolic regulation focus on concentration changes in the adenylates, PCr, and P_i as potential "driving functions" in activation of muscle ATP production and typically ignore the ATP demand side of the system. This turns out to be a mistake. While the classical control models may reasonably explain ATP production rates considered in isolation, it is clear that they are far "off the mark" for integrated control of ATP cycling between demand and supply. Let us consider the reasons for this conclusion by examining in detail a variety of muscle preparations.

GASTROCNEMIUS OF THE LABORATORY RAT

A particularly interesting set of data come from studies of Dudley et al., (1987) and Leijendekker et al., (1983) on the rat gastrocnemius. When free [ADP] is plotted as a function of O_2 flux, both data sets indicate similar patterns (Hochachka and Matheson, 1992). Data from trained rats or from hyperthyroid rats fall on curves that are clearly downward displaced relative to control conditions or to hypothyroid conditions. Furthermore, these curves are not analagous in form to Michaelis-Menten curves, which also is indicated by the observation that over a ten fold change in flux can be attained with only 2 to 2.5-fold change in free [ADP]. Although the data are not extensive enough to be certain, they are not readily accommodated by sigmoidal or Michaelis-Menten models.

BICEPS AND SOLEUS OF THE LABORATORY CAT

Somewhat more convincing results arise from recent noninvasive studies of phosphate metabolites during varying work rates in two different muscles of the cat (Kushmerick et al., 1992). The biceps is mainly fast-twitch, composed of about 25%

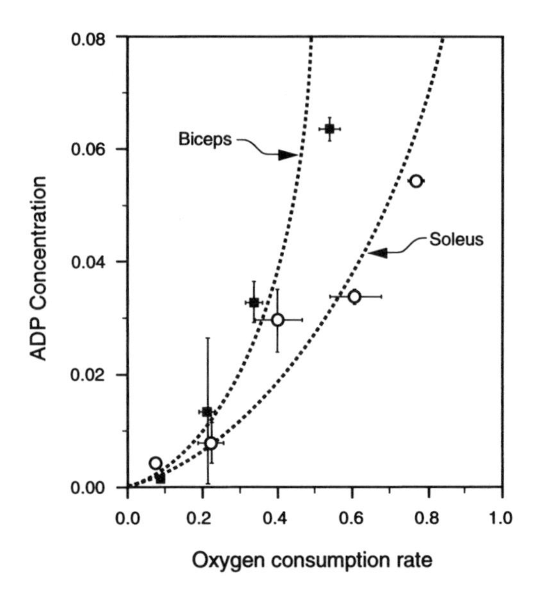

FIGURE 7–2. Plots of [ADP] in μmol g^{-1} vs. oxygen consumption rate (μmol O_2 g^{-1} min^{-1}) for cat biceps and soleus preparations at different stimulation frequencies (i.e., different work rates). Since the estimated maximum $\dot{V}O_2$ for these preparations is up to 12 times higher than these *ex vivo* rates, these data cover approximately the first 10% of the true metabolic scope for activity of these muscles. It is evident that if ADP were the only regulator of $\dot{V}O_2$, an approach to the true *in vivo* maximum rates would occur, with [ADP] rising essentially without limit. Data replotted, with *v* as the independent variable, from Kushmerick et al. (1992).

fast-twitch glycolytic oxidative or FOG-type fibers and 71% fast-twitch glycolytic or FG fibers. The soleus is mainly slow twitch. By varying the stimulation frequencies, the maximum *in vitro* metabolic rates observed were about 3 and 5 μmol ATP $g^{-1}min^{-1}$ for the biceps and soleus, respectively. Michaelis-Menten ADP saturation kinetics were assumed to be able to account adequately for these data; the observed maximum rates for the biceps were essentially the same as the calculated maximum, while the calculated maximum metabolic rates for the soleus were about 40% higher than the observed. So far so good. Unfortunately, the story does not end here. If these data are closely examined (or replotted with *v* as the independent parameter (Figure 7–2)), it becomes evident that modest changes in biceps metabolic rate correlate with large changes in [ADP], while larger changes in soleus metabolic rate correlate with lower changes in [ADP]. What is worse, the maximum observed rates of muscle metabolism achieved *ex vivo* are only a small fraction of maximum *in vivo* rates. Assuming that a 3-kg cat is similar to most mammals (working muscles equal to about one-third of body mass and accounting for 90% of $\dot{V}O_{2(max)}$, then it is easy to show that the maximum muscle metabolic rate is about 12 μmol O_2 g^{-1} min^{-1}, equivalent to about 72 μmol ATP g^{-1} min^{-1}. These estimates are about 25% higher than measurements made during trotting in cats; direct measurements of maximum O_2 consumption rates in galloping cats have not been achieved, because (not surprisingly to anyone who has ever owned one!) cats apparently do not like to gallop on a treadmill (C.R. Taylor, pers. comm.). Nevertheless, the above assumed $\dot{V}O_{2(max)}$ values are conservative and probably are not far off the mark; even lions some ten times heavier, achieve these kinds of metabolic rates (Taylor et al., 1981). This means that *in vivo* muscle metabolic rates for the cat biceps and the soleus are some 24 and

14 times higher, respectively, than the maximum observed *in vitro* rates. A realistic metabolic scope for cat muscles (maximum/resting metabolism) is perhaps 70-fold (similar to that found in many other mammals), not five-to-seven-fold found *in vitro*. Even during trotting, the average muscle metabolic rate would be about 54 times resting rates. In this event, the data of Kushmerick et al., (1992) at best cover the first 10% or so of the dynamic operating range of these systems—much too small a span to adequately test the validity of any metabolic control models. Put another way, over this range, such data can be made to fit almost any model of metabolic regulation. Even if the classical model proposed is accepted for the working range analyzed, it is hard to imagine how it could be expanded to cover the remaining 80 to 90% of the *in vivo* metabolic scope. Replotting the data in the *v* vs. *s* form discussed above clearly reveals that if extrapolated toward a realistic maximum metabolic rate, ADP concentrations would rise without limit (Figure 7–2). Thus, as in the data on rat muscles above, the behavior of the cat biceps and soleus cannot be readily accommodated by classical control models.

GRACILIS AND GASTROCNEMIUS OF THE LABORATORY DOG

In an elegant series of experiments on *in situ* dog gracili, Connett and his coworkers (1988, 1989, 1990) tried to sort out the metabolite signals contributing to the control of O_2 fluxes. However, cause-effect relationships were unclear. In their experimental conditions, as in others like them, PCr depletion and P_i accumulation are stoichiometric, and change in $[P_i]$ dominates change in the phosphorylation potential, which is proposed as a key metabolic regulator by these workers. However, it is now clear that through the work ranges achieved with these muscle preparations, which are inevitably far below *in vivo* maximum ranges, PCr and Cr can be largely removed from muscle cells without seriously impairing steady-state performance (see papers by Shoubridge and his coworkers, 1984; 1985; van Deursen et. al., 1993). The main impairments currently known would include initial phases of rest → work transitions and burst work per se, both for the same reasons (van Deursen et al., 1993). From these data taken together, it is more likely that phosphorylation potential reflects rather than ''drives'' dog gracilis ATP turnover rates.

Similar experiments on the isolated perfused dog gastrocnemius (Hogan et al., 1992; Arthur et al., 1992a) indicate that under normoxic conditions, the concentrations of putative regulators such as ADP and P_i change by a factor only slightly greater than 2, while ATP turnover rates increase by about 18-fold! What is even more difficult to accommodate by classical control models is the observation that putative regulators such as [ADP] change *much more under hypoxic conditions, even though ATP turnover rates decline relative to normoxic rates* (Figure 7–3).

GASTROCNEMIUS OF THE LABORATORY RABBIT

A recent ^{31}P-Nuclear Magnetic Resonance Spectroscopy (NMRS) study of nerve-stimulated rabbit limb muscles found good correlations between free [ADP] and tension time integral (TTI), considered to be an indirect measure of ATP flux (Nioka et al., 1992). Because the correlations held for several steady-state (normocapnic vs. hypercapnic) work conditions, the authors concluded that free [ADP] was probably the main determinant of ATP turnover rate. The study assumed Michaelis-Menten kinetics, but plots of TTI vs. free [ADP] do not go through the origin if reasonable resting metabolic rates (Kushmerick, 1985; Kushmerick et al., 1992; Hogan et al., 1992) are assumed. What is worse, 1.4-fold change in free [ADP] by the theory

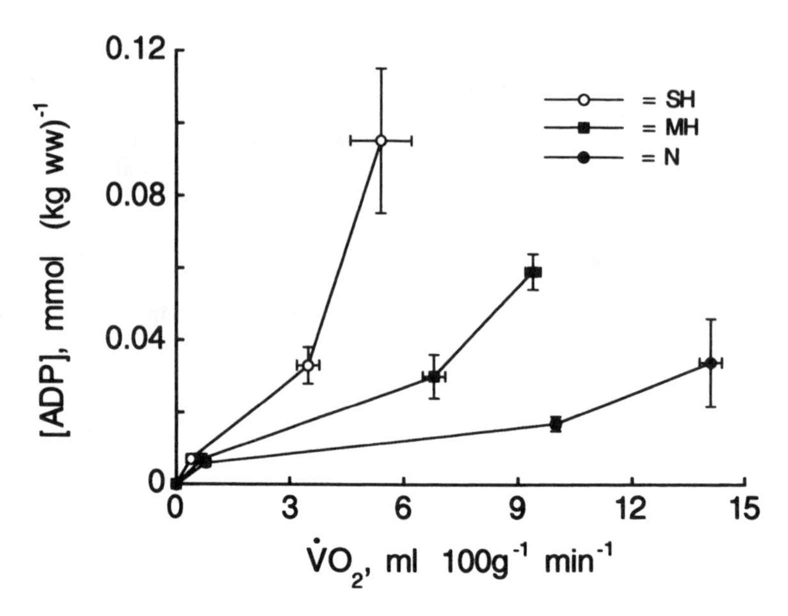

FIGURE 7–3. Plots of [ADP] as dependent variable in μmol kg^{-1} wet weight (ww) of muscle vs. $\dot{V}O_2$ in ml 100 kg^{-1} min^{-1} for dog muscle gastrocnemius under normoxia (dark circles), moderate hypoxia (dark squares), and severe hypoxia (open symbols). Under normoxic conditions, an 18-fold increase in O_2 consumption rates correlates with less than a three-fold change in [ADP]. This result is incompatible with standard metabolic control theory, invoking ADP as a critical metabolic regulator. What is even harder to explain in classical frameworks is the observation that under moderate and severe hypoxia, as the work rate and thus $\dot{V}O_2$ decline, [ADP] rises. Finally, the v vs. $[s]$ plots do not go through the origin, as would be required by Michaelis-Menten models of metabolic regulation (see Kushermick et al., 1992). Data replotted from Arthur et al. (1992a).

assumed would require 1.4-fold change in TTI; instead, up to a three-fold change was observed. Furthermore, replotting the data with free [ADP] as the dependent parameter indicates concave downward curvature, which is inconsistent with Michaelis-Menten kinetics.

BICEPS FEMORIS OF THE GREYHOUND—CANINE SUPER-ATHLETE

The above kind of discrepancy is also evident in a recent work on greyhound muscle metabolism during a simulated race (Dobson et al., 1988). Assuming normal resting metabolic rate for this muscle of about 1–2 μmol ATP g^{-1} min^{-1}, which is not unreasonable (Hogan et al., 1992), the greyhound biceps femoris sustained some 200 fold increase in ATP turnover rate while free [ADP] changed maximally by four-to-five-fold. Later, as mentioned above, in a stimulated dog muscle preparation under normoxic conditions, we observed about an 18 fold increase in O_2 flux with less than a 3-fold change in free ADP concentration (Hogan et al., 1992). These data again clearly are incompatible with Michaelis-Menten driven, or any other current, commonly popular models of metabolic regulation.

LEG MUSCLE OF THE THOROUGHBRED—EQUINE SUPER-ATHELETE

Thoroughbred horses, selected by man for high speed performance, have been studied in detail both under simulated race (Snow et al., 1985) and treadmill conditions (Rose et al., 1988). Assuming that about 30% of body mass and resting metabolic rate are due to skeletal muscle and that some 90% of $\dot{V}O_{2(max)}$ is due to skeletal muscle, the leg muscles of thoroughbreds sustain at least a 60-fold increase in ATP turnover rate (Rose et al., 1988). This is a low estimate since all muscles are assumed to be equally involved in locomotory work, while it is clear that muscles such as the gluteus maximus are especially important in supplying propulsive power. Nevertheless, these very large changes in metabolic rate are achieved with only 1.5- to 2-fold changes in PCr concentration and with no more than about 1.2 μmol/g change in ATP concentration; in fact, in the treadmill protocol, no change in [ATP] could be detected (Rose et al., 1988), implying similarly small changes in free ADP and AMP concentrations and in the total pool of phosphate metabolites usable in muscle work. Clearly these kinds of results could not be easily incorporated into Michaelis-Menten models of metabolic control.

CALF MUSCLE OF VARIABLY ADAPTED HUMANS

A recent noninvasive ^{31}P NMRS study of human gastrocnemius in high altitude adapted Andean natives compared to three differently trained subject groups (Hochachka et al., 1991; Matheson et al., 1991; McKenzie et al., 1991) yielded some insightful data. When the estimated rate of ATP turnover in the calf muscle during the exercise protocol for each of the groups (Matheson et al., 1991) is plotted as the independent parameter vs. free [ADP] and vs. $[P_i]$/, defined as the ratio of $[P_i]$/[PCr] + [Cr], the plots are similar in form and show that the four subject groups fall into three categories. Sedentary individuals sustain the lowest ATP turnover rates but the highest change in concentrations of ADP and P_i. Endurance trained (AER) and high altitude adapted Andean natives (AND) sustain the greatest ATP turnover rates but the least change in ADP and P_i concentrations, while power-trained subjects (PWR) show an intermediate pattern. Although the data available are compromised by the fact that mixed fiber beds were being interrogated, they nevertheless clearly indicate that none of the curves are consistent with Michaelis-Menten saturation kinetics. Finally, and again in contrast to expectations from classical control models, the highest ATP turnover rates were sustained with the lowest changes in metabolite concentrations. The muscles of Andean subjects, for example, sustained more than a 15-fold increase in ATP turnover rate (working metabolic rate compared to resting metabolic rate, RMR) but maximally about a five-fold increase in free [ADP].

SERVING A SMALL MUSCLE MASS WITH A LARGE CARDIAC OUTPUT

We have already mentioned the high metabolic rates achievable by the knee extensor, presumably because most of the cardiac output is able to serve the energetic needs of a relatively small muscle mass (Andersen and Saltin, 1985). Although free [ADP] for this muscle as a function of work rate was not determined, from the measurements of Graham et al., (1990), it is possible to get a good estimate of the total pool of phosphate bonds useful for muscle work. This parameter, termed P_e or the "potential energy" state of the cell, is defined as the sum of [PCr] + 2[ATP] + [ADP] and is proposed to be well correlated with muscle ATP turnover rates (Funk et al., 1990). Given standard RMR values of about 1 to 2 μmol ATP g^{-1} min^{-1} for mammalian muscle (Hogan et al., 1992), the knee extensor sustains close to two orders of magnitude increase in ATP turnover rate with only a two-fold change in P_e; moreover, the form of plots of P_e as a function of ATP turnover are inconsistent with Michaelis-Menten kinetics. Arthur et al. (1992a) observed similar patterns in plots of P_e vs. ATP turnover rates in studies of isolated dog gastrocnemius. Clearly, neither set of results is easily accommodated by classical control models.

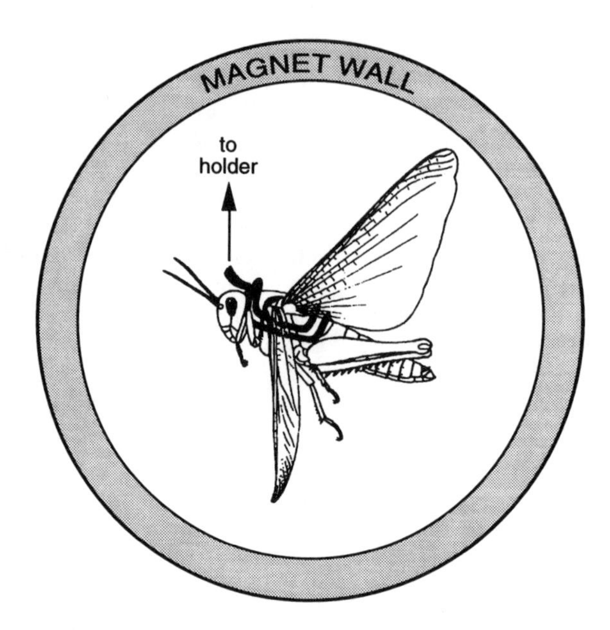

FIGURE 7–4. Real-time quantitative monitoring of the adenylates, of phosphagen, and of Pi in flight muscles of insects is now achievable with 31P Magnetic Resonance Spectroscopy (MRS). For such measurements, a custom-designed radiofrequency coil is worn like a small harness and positioned over the flight muscles of the locust. The coil is attached to a holder, and the entire set-up is then positioned within the homogeneous magnetic field of a wide-bore magnet. Advantage was taken of a well-known behavioral characteristic of many insects: the locust would not fly if presented with a platform (a small plastic rod) for its feet. On removal of the platform, a flight reflex would be activated. Because the insect was tethered, the exercise was a fixed-flight analogue of forward flight. On rest → flight transition, the ATP turnover rate of locust flight muscle is known to increase by over 600 times. However, Wegener et al., (1991) found only a 50% decline in phosphagen. On the assumption of a 7% drop in ATP, they concluded that these muscles may have sustained a maximum five-fold increase in ADP concentration, about 100-fold less than would be required for it to be the sole signal for accelerating ATP turnover rates. Modified after Wegener et al. (1992).

FLIGHT MUSCLE OF INSECTS—CHAMPION ANIMAL ATHLETES

A particularly dramatic example of ATP demand (as independent parameter) vs. change in [pathway intermediates] is available from studies of various intermediates as well as the adenylates in insect flight muscle (Rowan and Newsholme, 1979; Sacktor and coworkers, 1966a,b). These clearly show that the relationships between flux and pathway [substrates] do not fit Michaelis-Menten models. They likewise show that several hundred-fold (!) increase in ATP turnover rate is sustainable with much more modest—and often immeasureable—change in substrate concentrations. This large discrepancy is also noted by others in the literature on adenylates in flight muscles of insects. One of the more interesting of such recent studies involves real time MRS monitoring of ATP, phosphagen, and P_i in the flight muscles of locusts during rest and during steady-state forward flight (Wegener et al., 1991). This was made possible by special custom-designed uptake coils that were worn by the insect; coil and all were then placed in the homogeneous magnetic center of a spectrometer (Figure 7–4). The result was a high-tech version of earlier, more invasive studies: despite the well-established approximately 600-fold increase in ATP turnover rate on rest → work transition, Wegener et al., (1991) were hard pressed to see comparably large metabolite concentration changes. Instead, these *in vivo* studies confirm that during flight in the locust, phosphagen concentrations in flight muscle decline by only 50% of resting values and P_i increases essentially stoichiometrically. ATP and pHi are stable, and on the assumption that arginine phosphokinase (APK) remains in equilibrium, *these data mean that one of the most active muscle machines in nature is turned on to operate at 600 times resting ATP turnover rates with but a two-fold change in [ADP].* Even if is assumed that [ATP] drops by 7% (not experimentally demonstrable), these muscles sustain at most a five-fold increase in [ADP]. This is still less than 1/100th the change that would be required by classical control theory for ADP to ''drive'' the very high ATP turnover rates that occur in this muscle during flight. Yet, the authors conclude that their data supply evidence in favor of the classical ADP-drive model of metabolic regulation (!), which we take as an interesting illustration of the power of dogma to constrain the creativity and intellect of our colleagues.

THERMALLY DRIVEN CHANGE IN ATP TURNOVER RATES

Analogous pictures emerge when we closely examine situations in which temperature, rather than change in work rate per se, influences muscle ATP turnover rates. Two recent studies have come to our attention. The first focuses on the rattler muscle of the rattle-snake, which appears to have some interesting adaptations designed for sustained high frequency rattling (up to 60 min or so under some conditions); to this end, mitochondria and SR volume densities have both been up-regulated (see Chapter 8). Recent MRS studies of ATP, phosphagen, and P_i in these muscle cells yield intriguing results: with the onset of rattling, PCr concentrations fall to the same steady-state values at all temperatures studied. Because of the temperature coefficient of metabolism and work, the ATP turnover rates at the lower and upper thermal extremes in such studies differ by at least a factor of five-fold. However, no temperature-dependent change in [ADP], the putative regulator of ATP turnover favored by these workers, is observed under these conditions (Linstedt, S. and K. Conelly, pers. comm.).

A second study involving thermally driven change in metabolic rate focuses on the scapular muscles of hibernating squirrels (McArthur et al., 1990). In this case, phosphagen, ATP, and P_i were monitored by MRS techniques as the squirrel aroused from hibernation and as the temperature of its shoulder muscles increased by 20° C. As in other systems (Hochachka and Somero, 1984), it is not unreasonable to assume about a four- to six-fold difference in mass-specific metabolic rates of these muscles over the thermal range studied. Although metabolic regulation was not the reason for this research, it is noteworthy that the 31P spectra obtained as the shoulder muscles warmed were constant and independent of temperature (McArthur, M.D., pers. comm.). This means that the concentrations of putative regulators such as ADP also remain largely independent of temperature, despite the fact that ATP turnover rates change by four- to six-fold under these conditions. It thus would seem that neither the rattler muscles nor the muscles of the arousing hibernator can easily be fitted to classical paradigms of metabolic regulation.

PATHWAY INTERMEDIATES AND THE LATENT ENZYME CONCEPT

To this point, our analysis has focused mainly upon so-called ''high energy'' phosphate metabolites and P_i, but analogous difficulties arise when other pathway intermediates are considered. At near-maximum running exercise in the rat, the concentrations of glycolytic intermediates in working skeletal muscles change by about two-fold, some going up, others going down; at the same time, rates of ATP turnover may increase by two orders of magnitude (Hochachka et al., 1991). In fish white muscle of several species, glycolytic intermediates again change by only two- to three-fold, but glycolytic flux may change by several hundred-fold (see Moyes et al., 1992, for literature in this area). Similarly, insect flight muscles sustain very large increases in glycolytic flux to the level of pyruvate, which is then fully oxidized by the mitochondria, with minute perturbations in [pathway intermediates].

In this regard, analysis of a specific step in metabolism, such as that catalyzed by pyruvate kinase, is instructive. In vertebrate muscles sustaining orders of magnitude increase in glycolytic flux during burst (anaerobic) work, [PEP] changes hardly at all, [ADP] increases two- to three-fold, [ATP] declines modestly (about 30%), while [pyruvate] increases by two- to three-fold. Almost an identical profile typifies the pyruvate kinase step in an essentially obligatorily aerobic carbohydrate metabolism in insect thoracic muscles during flight. These data are inconsistent with models of glycolysis assuming that all enzymes in the pathway operate at near-equilibrium and that for all there is an appropriate change in [substrate]/[product] ratios to achieve the large rate changes observed. The most powerful of such models (Betts and Srivastava, 1991), while philosophically attractive because it assumes metabolite channeling along much or all of the pathway, requires that the $K_{eq} = 1$ for each step in the pathway and that flux acceleration and net flux in steady state are driven by [substrate] increases or [product] decreases. Neither the direction nor magnitude of changes in substrates and products for the PK reaction in the above (anaerobic or aerobic) glycolytic systems is compatible with the Betts/Srivastava model. Thus, it appears that no convolution of these changes in substrate and product concentrations alone could cause the observed large order change in flux through this kind of step in metabolism during muscle work.

For allosteric enzymes, or for enzymes under phosphorylation-dephosphorylation control, this large discrepancy can be explained by assuming large modulator effects on enzyme-substrate affinities, on maximum velocities, or on both. But this does not apply for many (indeed most) metabolic enzymes that display Michaelis-Menten kinetics with no known modulator effects. We argue that the only way that flux changes of 100-fold or more can occur for simple catalysts with modest (two- to three-fold) change in [pathway intermediates] is through increasing the effective concentration of pathway enzymes. In most and maybe all muscles for which good data are available, there seems to be a requirement in the resting state for the existence of a latent pool of enzymes—*a pool of enzymes that are inaccessible to (or nonreactive with) their substrates.* Muscle activation, by this interpretation, would involve an effective unmasking or release of enzymes from this latent pool so as to allow large changes in flux while keeping near-normal (near-resting) concentrations of pathway intermediates. In terms of the classical enzyme velocity equation, $V_{max} = e_o \times k_{cat}$ (where e_o is the enzyme molar concentration and k_{cat} is the turnover number per catalytic site), *this model postulates that e_o regulation dominates large-scale changes in ATP turnover rates, while concentration changes in substrates, products, and modulators are the fine tuning mechanisms that help to account for the inordinate precision of flux control.* This hypothesis (Hochachka and Matheson, 1992) of how metabolic pathways are regulated is similar to that for the *in vivo* regulation of Na^+K^+ ATPase and CPK during electric organ discharge, which is reviewed by Blum et al., (1991), and for the *in vivo* regulation of myosin ATPase. In the resting electric organ (i.e., in the absence of appropriate activation signals), the Na^+K^+ ATPase is, in effect, a

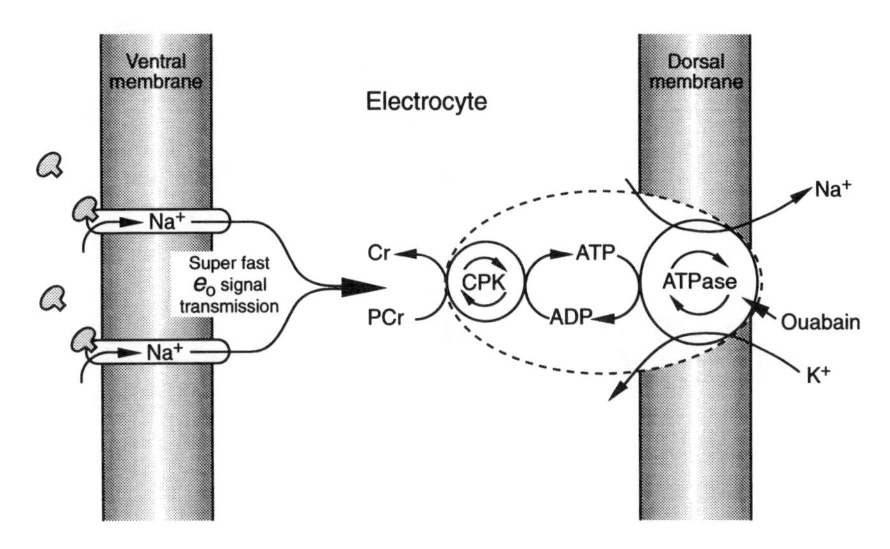

FIGURE 7–5. Model of control of ATP turnover in a modified muscle, the electric organ of electric fish. Acetylcholine is thought to activate the acetylcholine receptor, a signal that is almost instantaneously transmitted to CPK and to Na^+K^+ ATPase. The activation process takes a maximum of 300 ms and involves a 2000-fold activation of the ATPase; CPK activity manages to pace this enormous activation. Neither metabolite nor ion concentration changes are anywhere near large enough to account (i) for the "flare up" rate or (ii) the absolute activation level. The only mechanism that seems to be able to account for this behavior assumes unmasking of both CPK and Na^+K^+ ATPase catalytic capacities, which are largely latent in the resting state. Data summarized from Blum et al. (1990; 1991).

latent enzyme. In the case of resting muscle (i.e., in the absence of appropriate Ca^{++} signals), myosin is, in effect, a latent ATPase. In both cases, a full and enormous catalytic potential is released when and only if properly acted upon from the outside. A possible model of the kind of control system that is required is summarized for Na^+K^+ ATPase in Figure 7–5. We consider that an analogous process may be occurring in metabolic enzyme activation that may have been previously overlooked, perhaps because of dogma dominating the field (that effective enzyme concentration is the same under rest and work conditions) or because of the destructive methods normally used in enzyme studies.

It is instructive that a similar control process (unmasking of latent catalytic potential) is already known for the regulation of red blood cell glycolysis. The point of departure for this concept, developed by Low and his colleagues (1990), is the observation that several glycolytic enzymes (such as glyceraldehyde-3-phosphate dehydrogenase, GAPDH, used in these studies) and other proteins specifically associate with the N-terminus tail of the cytoplasmic domain of band 3 protein (Figure 7–6). Band 3 protein occurs in high concentrations in the membranes of red cells, as well as in many other cells, and plays several known roles. One of these is to serve as the chloride-bicarbonate exchanger. Another is to regulate glycolysis. When bound to band 3 protein, GAPDH is catalytically inactive, while upon release, its full catalytic activity is unmasked. In terms of catalysis, in the bound state, the effective concentration of GAPDH is zero. Studies with resealed red blood cells show that GAPDH binding is controlled by phosphorylation of tyrosine 8 and tyrosine 21 within the GAPDH binding site on band 3; phosphorylation apparently obstructs enzyme binding, which leads to GAPDH activation. Modeling indicates that GAPDH interacts

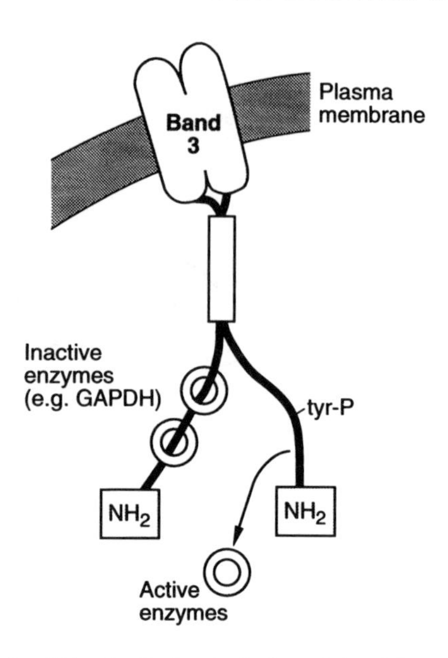

FIGURE 7–6. Control of red blood cell glycolysis through band 3 protein binding and thus inactivation of glycolytic enzymes such as GAPDH. GAPDH and other glycolytic enzymes are thought to bind to the N-terminal tail of band 3 protein; in this state, catalytic activity is fully masked. PTK catalyzed phosphorylation of two tyrosine residues on the N-terminal tail prevents binding and releases the full catalytic activities of bound enzymes. Because the turnover number of glycolytic enzymes is characteristically extremely high, the impact of releasing a relatively small amount of enzyme is enormous. (Modified from Low et al. (1990) and Harrison et al. (1991).)

with band 3 like a bead on a string, and the interaction is sterically blocked when the band 3 tail tyrosines 8 and 21 are phosphorylated. Protein tyrosine kinase (PTK) catalyzes band 3 phosphorylation and itself is thought to be regulated by receptors located on the plasma membrane, because agents that activate PTK activate glycolysis, and PTK inhibitors do the reverse (Harrison et al., 1991; Low et al., 1990). It is important to stress that, at this time, such regulatory systems for skeletal muscle metabolic activation are not known and are not considered likely, simply because they may be too sluggish for the large scale rest → work transitions typical of muscle. Yet, exactly analogous regulation (controlling the pool of catalytically active enzymes) well accounts for increasing metabolic flux by orders of magnitude while sustaining negligible change in concentrations of pathway intermediates.

EXOGENOUS CONTROL OF ENERGY COUPLING

From the above analysis, one conclusion seems inescapable: most current efforts at unravelling mechanisms regulating ATP turnover rates, directed as they are at mainstream pathway intermediates (such as the adenylates, phosphagen, P_i, pH, and so forth), are essentially misdirected. From the data considered, it is evident that changes in intermediates of ATP utilization and synthesis pathways are modest and in themselves cannot account for overall flux adjustments. What seems to be required, therefore, is insight into the nature or identity of the (probably exogenous) regulators of ATP turnover. This need was clearly evident in studies of insect glycolytic control requirements under aerobic conditions by the mid-1960s (Sacktor and coworkers,

1966a,b) and was especially clearly enunciated in studies of control requirements for insect muscle mitochondrial metabolism in 1979 by Rowan and Newsholme. The latter focused on locust flight muscle, which can attain aerobic metabolic rates of 600 to 800 μmol ATP per gram^{-1} minute^{-1}; assuming usual RMR values for muscle, this rate approaches three orders of magnitude up-regulation of flux with minimal concentration change in Krebs cycle intermediates. Both of these earlier studies recognized implicitly or explicitly the main conclusion of this chapter, namely, that key control elements of energy consuming and of energy yielding pathways *must lie essentially external to the pathways per se* since the concentrations of the intermediates do not change by a large enough factor to cause the flux change directly. What is more, all of the above data seem to require *simultaneous control of both the ATP demand and the ATP supply arms of the ATP bioenergetic cycle.* Otherwise, large change in work rate would not be possible with steady concentrations of adenylate intermediates. Although the nature of such controlling elements as well as their mechanisms of action at this time are unknown, some interesting clues arise from recent studies in two different lines of research, both pointing to the possibility of an important regulatory role for O_2 and O_2 sensing systems, at least for aerobic metabolism.

OXYGEN SENSING IN REGULATION OF ATP TURNOVER

The first of these two lines of research initially focused primarily on the hypoxia mediation of erythropoitin (EPO) synthesis but simultaneously may have exposed a universal O_2 sensing system. Earlier studies had already confirmed the occurrence of an O_2 sensor, possibly a heme protein, which in its deoxy form triggers an activation pathway culminating in increased transcription of the EPO gene. More recent studies suggest that the occurrence of an O_2 sensing system and of an enhancer sequence regulating gene transcription are much more widespread than had been previously believed. These studies, based on the development of a molecular reporter system allowing the experimenter to rapidly scan many cell types for *enhancer-specific hypoxia sensitivity, indicate that O_2 sensitive gene expression is common in about 90% of the cell lines tested.* Under hypoxia conditions (Figure 7–7), the signal (reduced O_2) mediates the production of hypoxia-inducible factor (HIF-l), which binds to a specific "hypoxia induction site" on DNA and accelerates transcription of O_2-regulated genes (including the EPO gene and the genes for at least some glycolytic enzymes and the GLUT1 isoform of the glucose transporter (Bashan et al., 1992)). Although current studies clarify the function of this complex O_2 sensing system only in hypoxia regulation of EPO production, by serendipity they may have revealed a universal molecular pathway for sensing declining O_2 availability at the cell level (Wang and Semenza, 1993; Maxwell et al., 1993; Eckhardt et al., 1993).

The second of the above two lines of research was designed to tease out the roles played by O_2 availability in control of ATP turnover during muscle work under normoxia, moderate hypoxia, and extreme hypoxia. A dog muscle preparation was used in order to bring the system as much under the experimenters' control as possible (Hogan et al., 1992; Arthur et al., 1992a). These studies confirmed that, as argued above, none of the usual putative regulatory metabolites (such as ADP or P_i) can account for the changes in ATP turnover rates observed—*except possibly for oxygen itself.* The roles of O_2 are complex and are best explained by looking at ATP demand (proportional to ATPase) first, then at the metabolic response (Figure 7–8). Under normoxia, increasing stimulation frequency from 0.5 to 1 Hz leads to increasing ATPase activity (increasing work), which, of course, is expected. What is somewhat surprising is that moderate hypoxia at 0.5 Hz brings about a *regulated* decline in

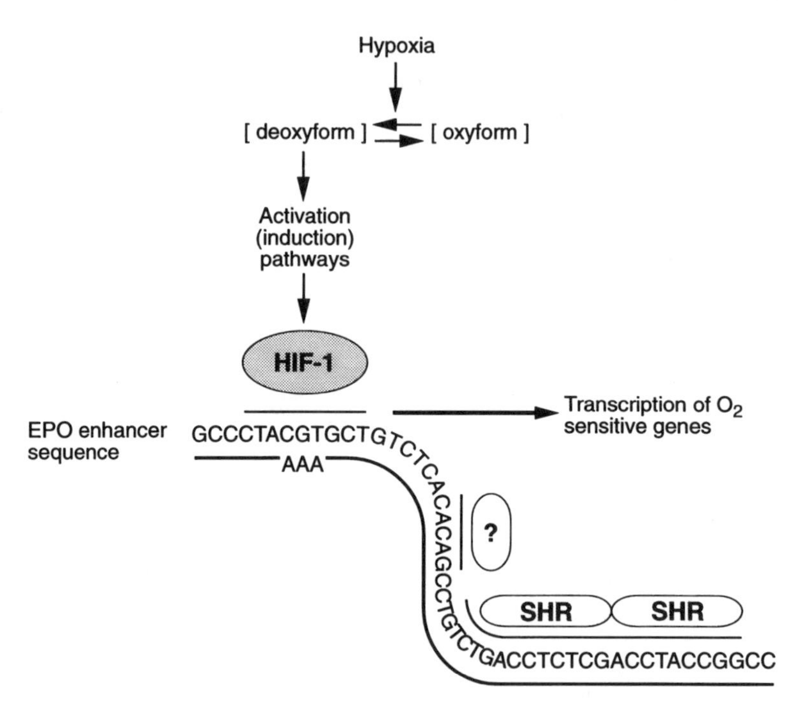

FIGURE 7–7. Diagrammatic structure of the hypoxia-inducible enhancer of the EPO gene. Under hypoxic conditions, the fractional deoxygenation of the putative O_2 sensor increases and leads to the induction of HIF-1 (hypoxia inducible factor-1). Only a part of the 50-nucleotide sequence of the EPO enhancer forms the HIF-1 responsive (binding) element adjacent to regions responsive to other signals (SHR). Transcription of reporter genes (e.g., the gene for chloamphenicol acetyltransferase) containing the EPO gene enhancer sequence is induced by hypoxia in many (and possibly all) cell types. Since mutations that eliminate HIF-1 binding eliminate hypoxia induction, this factor is considered pivotal in this pathway of O_2 sensitivity. Modified from Wang and Semenza (1993).

ATPase activity. We know this is a hypoxia-regulated rate because if the muscle is stimulated harder (at 1 Hz), its ATPase activity reaches that characteristic of normoxic conditions at 0.5 Hz. A similar if more extreme situation occurs at extreme hypoxia. In strictly empirical terms, what these data mean is that the ATP demand during mechanical work is *set by the ATPases but regulated at least in part by oxygen availability,* and the question arises of how the ATP supply side of the system responds.

ATP supply mechanisms respond to increasing stimulation frequency in this dog muscle preparation with an 18-fold increase in O_2 consumption at 1 Hz under normoxic conditions. Again (Figure 7–9), at 0.5 Hz, moderate hypoxia brings about a *regulated* decline in ATP synthesis rate. As before, we know this is a regulated rate because at double the stimulation frequency, the metabolic rate in hypoxia increases to normoxic 0.5 Hz values. In empirical terms, these data show that the ATP supply side of the system is also regulated by O_2 availability. What is more, the regulation is coordinated (Figure 7–10), with percent change in ATP demand being almost perfectly balanced by percent change in ATP supply (Hochachka and Matheson, 1993). Because of this coordination, ATP turnover rate is a linear function of O_2 delivery in this preparation (Figures 7–11 and 7–12); here, as in many other systems studied by other workers, *the only currently known "signal" that is large enough and that*

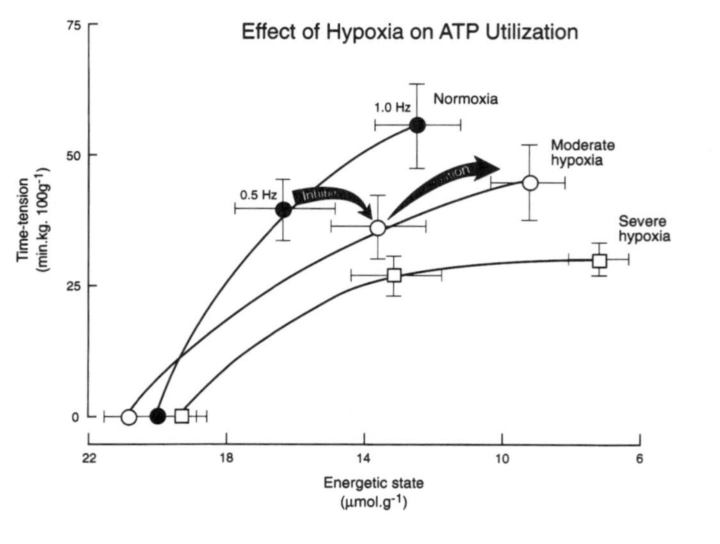

FIGURE 7–8. Effect of hypoxia on muscle ATP utilization (proportional to time-tension integral) as a function of the utilizable (high energy) phosphate. Data in each case shown for resting conditions (high energetic state), for 0.5 Hz stimulation (intermediate energetic state), and for 1 Hz stimulation (lowest energetic state). Note that under normoxia, increasing stimulation frequency (decreasing energetic state) increases the work output. At any given condition, hypoxia decreases contraction-linked ATPase activity (shown by arrow moving from normoxia to moderate hypoxia). This is a regulated response because increasing stimulation from 0.5 to 1.0 Hz increases ATPase to near normoxic levels (shown for moderate hypoxic conditions by arrow moving from intermediate to low energetic state). These kinds of data are consistent with ATP demand being set by stimulation frequency but being modulated by O_2 availability. (Modified from Arthur et al. (1992a).)

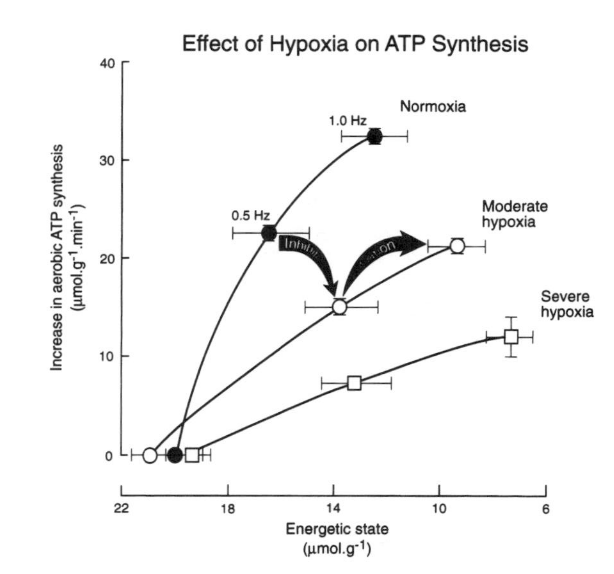

FIGURE 7–9. Effect of hypoxia on ATP synthesis in dog muscle gastrocnemius at various work levels. Data are shown for resting conditions (highest energetic state), for 0.5 Hz stimulation (intermediate energetic state), and for 1 Hz stimulation (lowest energetic state). Under normoxia, increasing stimulation (decreasing energetic state) correlates with increasing ATP synthesis rates. On imposition of hypoxia at 0.5 Hz, ATP synthesis rates decline (shown by arrow going from normoxia to moderate hypoxia). That this, too, is a regulated response is indicated by the fact that at higher stimulation frequency (1 Hz), ATP synthesis rates increase to normoxic range, despite the sustained hypoxia. These data are consistent with ATP synthesis rates being set by the ATP demand but being modulated by O_2 availability. (Modified from Arthur et al. (1992a).)

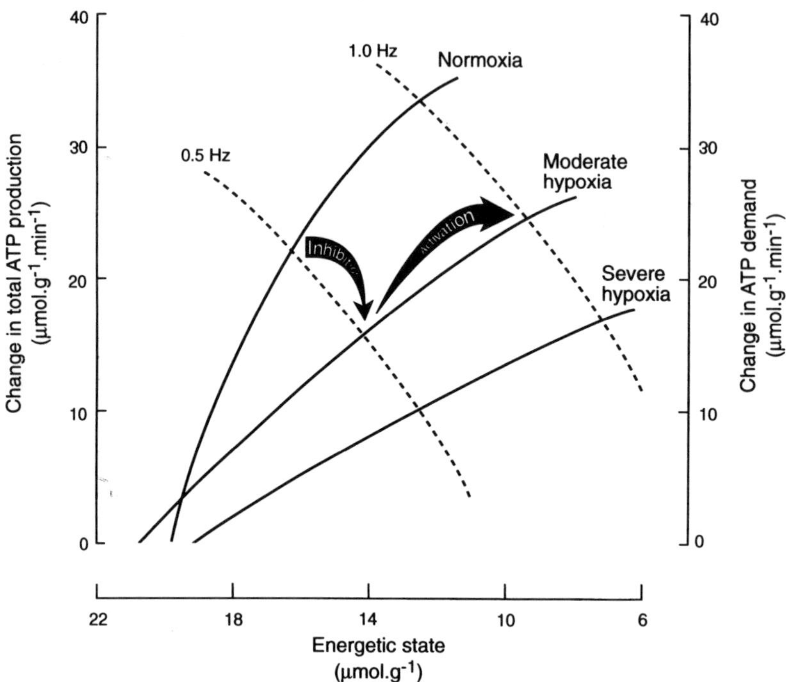

FIGURE 7–10. A composite summary of the effects of O_2 availability on ATP utilization rates (dashed curves) and on ATP synthesis rates (solid curves) at two different stimulation frequencies. As in figures 7–7 and 7–8, it is evident that at 0.5 Hz, transition to hypoxia decreases both ATP demand and ATP supply pathways (shown by arrow going from normoxia to moderate hypoxia at 0.5 Hz). This is taken to be a regulated response because increasing stimulation frequency can drive ATP turnover rates back to normoxic values, despite sustained hypoxia (shown by arrow for data at moderate hypoxia going from 0.5 Hz to 1 Hz). The maximum ATP turnover rate at any given condition is where the two curves intersect. These data are consistent with O_2 regulatory effects on both ATP demand and ATP supply pathways. (Modified summary from Arthur et al. (1992a).)

is directly proportional to the observed change in ATP turnover is O_2 availability. This also may be the explanation for the perplexing observations on the lack of PCr resynthesis during recovery under ischemic conditions in human muscles. For an experimental period of up to 5 min of ischemic recovery, the concentrations of PCr, P_i, ATP, ADP, AMP, and H^+ remain at the levels attained at the end of the work period. Despite appropriate conditions (high concentrations of putative positive modulators) for very high rates of glycolytic ATP and thus PCr synthesis, no such fluxes are observed until perfusion is reestablished (Quistorff et al., 1992). No matter what else these data indicate, they imply (i) that classical control models of glycolysis are inadequate to account for its paradoxical suppression under these conditions and (ii) that the recovery ATP synthesis rates are a function of O_2 availability.

In all these kinds of systems, however, O_2 regulation is not simply mediated by O_2 as a substrate for mitochondrial metabolism. In the dog muscle preparation, as in many other comparable systems where similar effects have been observed (Thurman et al., 1993), O_2 regulation cuts in at concentrations that are considered to be fully saturating for mitochondria (in other words, O_2 is influencing ATP turnover rates at

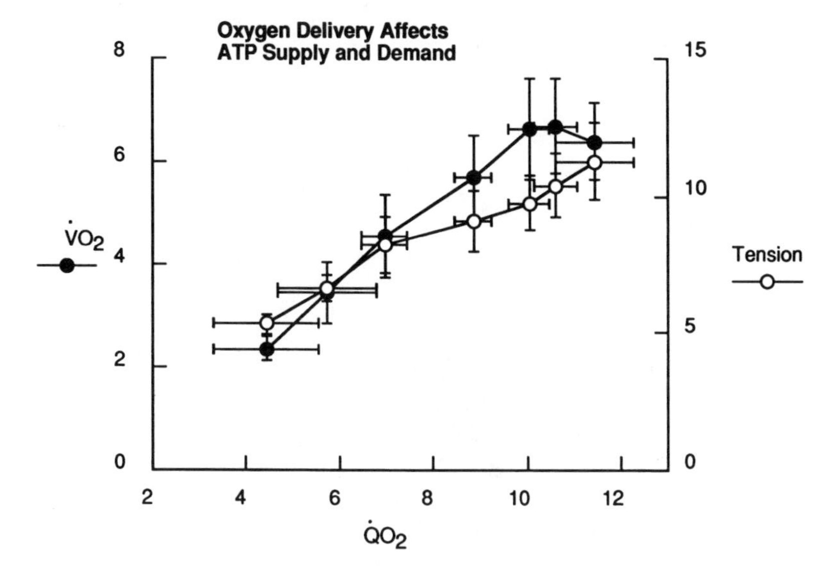

FIGURE 7–11. Changing O_2 delivery ($\dot{Q}O_2$ in ml O_2 per 100 gram per minute by imposing moderate hypoxia affects both ATP supply pathways (measured as $\dot{V}O_2$) and ATP demand pathways (measured as tension-time integral). Conditions as in Hogan et al. (1992) and Arthur et al. (1992). From Hogan, M. and Arthur, P. (unpublished data).

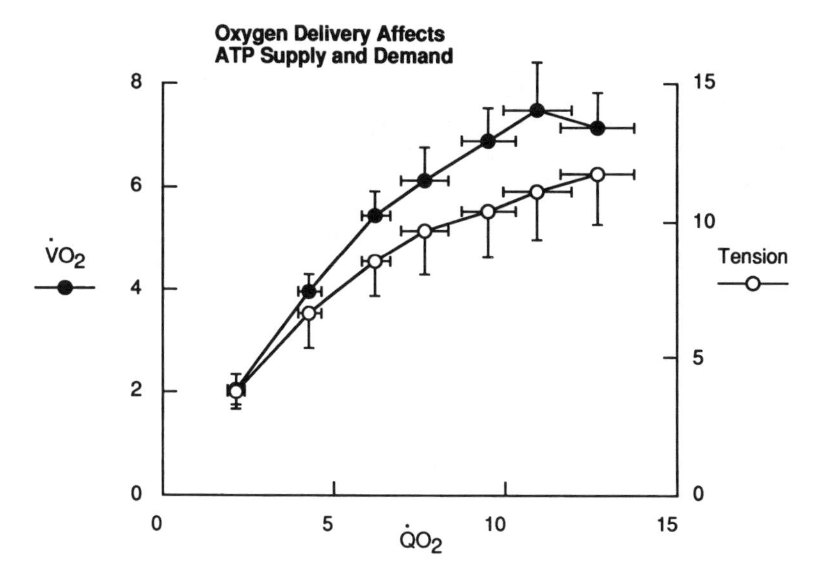

FIGURE 7–12. Changing O_2 delivery ($\dot{Q}O_2$ in ml O_2 per 100 gram per minute by ischemia affects both ATP supply pathways (measured as $\dot{V}O_2$) and ATP demand pathways (measured as tension-time integrals). Conditions as in Hogan et al. (1992) and in Arthur et al. (1992). From Arthur, P. and Hogan, M. (unpublished data).

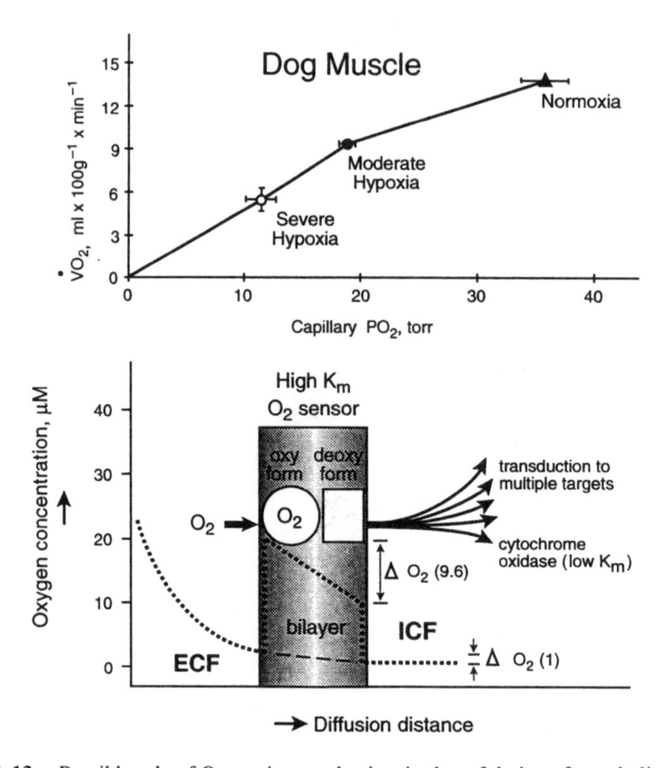

FIGURE 7–13. Possible role of O_2 sensing mechanism in the refulation of metabolism at varying O_2 availability. *Upper panel:* Changing O_2 consumption rates ($\dot{V}O_2$ in ml O_2 100 g^{-1} min^{-1}) as a function of capillary O_2 concentrations in working dog gastrocnemium (Hogan et al., 1992). The reader should realize that the relationship is between $\dot{V}O_2$ and extracellular PO_2; under the working conditins described, intracellular PO_2 is thought to be fully saturating to mitochondrial metabolism (Arthur et al., 1992). A diagrammmmatic and hypothetical interpretation of how an O_2 sensor might be involved in regulating ATP turnover rates is presented in the *lower panel*. The problem of sensing O_2 for a working cell is complex since the higher solubility of O_2 in the lipid bilayer of membranes creates higher O_2 concentrations and gradients. Under the conditions described, the difference in concentration across the lipid bilayer is some 9.6 times greater than the difference between the two aqueous compartments. Because of large solubility differences, $[O_2]$ in aqueous solution equal to 1.18×10^{-6} M Ton^{-1} corresponds to $[O_2]$ in the lipid bilayer equal to about 1.18×10^{-5} M $torr^{-1}$. What part of this complex situation is sensed by cells is unclear at this time; in this diagrammatic model, the sensor is positioned in the membrane (where the external to internal gradient is maximized) to emphasize the correltaiton between extracellular O_2 and $\dot{V}O_2$ (shown in upper panel). $[O_2]$ values given in lower panel are arbitrary, chosen only to qualitatively illustrate the situation.

concentrations well above the K_m for mitochondrial metabolism per se). The most parsimonious model to accommodate such striking and often-observed dependence of ATP turnover on O_2 availabilty is that of *a single O_2 sensor, which displays a much lower apparent O_2 affinity than typical of mitochondria and which (directly or indirectly) modulates both ATP demand and ATP supply pathways in a simultaneous coordinated way* (Figure 7–13).

Instead of being linked to HIF-1 induction, the O_2 sensor in this case would have to be linked to an intracellular, long-range, multitarget activation pathway. In outline, such a model can be summarized as follows: ischemia or hypoxia → declining O_2 delivery to cells → increasing [deoxy form]/[oxy form] of the sensor → → masking

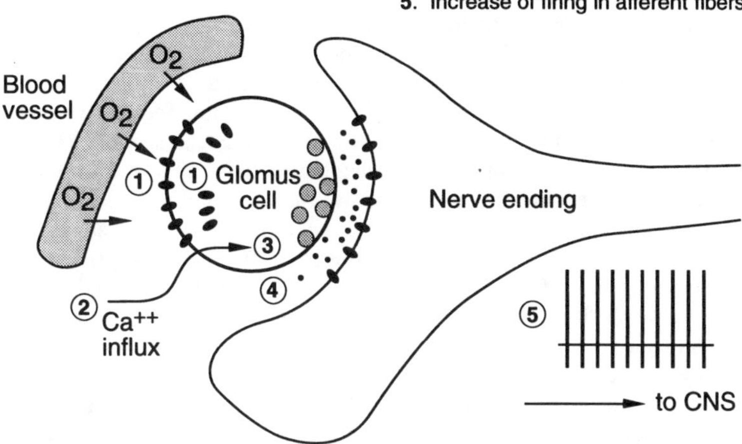

FIGURE 7–14. Diagrammatic summary of an O_2 sensing system in the glomus cell utilized during chemoreception. (1) O_2 sensor, which would be located on the cell membrane if it were an O_2 sensitive K^+ channel, but which would be located within the cell according to competing models of O_2 sensing; (2) hypoxia activation depolarizes the glomus cell, leading to a Ca^{++} influx and (3) a rise in intracellular [Ca^{++}]; (4) transmitters are released as discussed in Chapter 2; (5) the released transmitter activates the postsynaptic membrane and increases the firing rate. Firing rate is proportional to hypoxia: the lower the oxygen availability the higher the firing rate. Modified after Lopez-Barneo et al. (1993).

of many catalytic potentials → down regulation of ATPases and of ATP producing pathways. The reverse scenario is postulated for conditions of improved O_2 delivery, with increasing oxygenation state of the putative O_2 sensor and with enzyme unmasking causing ATP turnover rates to accelerate. From what is currently known, there are two minimal requirements that this kind of model would need to be explained: (i) it would need to account for the simultaneous activation of both ATP demand and ATP supply pathways during activation of muscle work and O_2 delivery, and (ii) it would need to account for intracellular long range activation; i.e., for a mechanism to "spread" the signal to all parts of the cell at once.

O_2 SENSING—POSSIBLE PATHWAYS AND MECHANISMS

It should be emphasized that the problem of O_2 sensing has already been examined by physiologists and biochemists in the past, mainly in the context of carotid body chemoreception (Figure 7–14). At a recent International Union of Physiological Sciences conference in Glasgow, three O_2 sensing models were favored: a high K_m mitochondrial model, a nitric oxide (NO)-based model, and an O_2-derived free radical model; other evidence has pointed in the direction of a unique outwardly poised, oxygen-sensitive K^+ channel (Lopez-Barneo et al., 1993). The erythropoitin (EPO) field, in contrast, assumes a heme protein as the O_2 sensor. These ideas are reviewed elsewhere (Hochachka and Monge, 1994). For our purposes in the present context, it is sufficient to emphasize that experimental probing into the potential role of O_2 sensing in regulation of ATP turnover can be organized into direct or indirect models. The first and most direct assumes an O_2 sensing system analogous or homologous to that involved in EPO regulation. Known to be widely and possibly universally distributed (Maxwell et al., 1993), this kind of O_2 sensing system has the experimentally advantageous feature of being modifiable by specific agents; for example, hypoxia and cobalt have similar but nonadditive effects.

A second framework for O_2 sensing in regulation of ATP turnover assumes an indirect pathway that is based on recent experimental data showing that nitric oxide under physiological conditions inhibits both ATP production, indirectly measured as O_2 uptake rates, and ATP utilization, measured as mechanical work (King et al., 1993). Based on these kinds of observations, the pathway that is in the process of being evaluated can be summarized as follows: ischemia and/or hypoxia → reduced O_2 delivery → increasing intracellular Ca^{++}/calmodulin concentrations → nitric oxide synthase (NOS) activation → increasing NO flux → several effects including simultaneous inhibition of ATPases and ATP synthesis pathways. Alternatively, increased NO availability could act via guanylate cyclase activation and cGMP to down-regulate energy demand and energy supply pathways, as is thought to occur in vascular smooth muscle (Knowles and Moncada, 1993; Toda and Okamura, 1992). The reverse processes are postulated to regulate rate transitions during improved perfusion and oxygenation characterizing muscle activation. Preliminary evidence favoring this indirect O_2 response pathway comes from adminstration of agonists or antagonists that specifically block or enhance NO production; perfusion, O_2 consumption, and work output are all proportionately up- or down-regulated (Sun and Reis, 1992; King et al., 1993). Although there are serious difficulties with this indirect model of the role of O_2 in metabolic regulation (NO has a notably finite life time and is destroyed on encounter with hemoglobin), it retains the experimentally advantageous features of being predictive and testable. Hence, it should be a simple matter of time and a bit more work for researchers to either refute the hypothesis or place it on an even firmer experimental foundation.

Neither of the above two approaches, unfortunately, address the issue of a long range intracellular "signaling" system to instantaneously "spread" the activation signal to all parts of the muscle cell. A quick glance at popular texts or recent reviews will make it clear that most current approaches to intracellular signaling are dominated these days by the second messenger concept. It may well be that some superactive second messenger system is in fact pivotal in turning on muscle ATP turnover processes; Ca^{++} has already been mentioned as possibly playing such a pivotal regulatory role. The limitation of all such second messenger frameworks is the requirement for highly specific intracellular targets. *The biological reality is that every protein in every step in ATP demand and ATP supply pathways must be almost simultaneously activated,* and it is hard to visualize a second messenger hitting that many targets that fast. Hence, something is missing in this classical view of muscle activation. We consider that the missing element in all current models of control of ATP turnover may be recognition of the nature of the intracellular milieu itself.

CONTROLLING THE PHYSICAL STATE OF ICF

Our insight into the possible importance of this arose by serendipity, when we turned to barnacle muscle cells to try to get estimates of true resting concentrations of metabolites such as ADP (Arthur, P. and Hochachka, P.W., unpublished studies).

We chose the giant barnacle because it is well known to have giant muscle cells (1 to 3 mm in diameter and 10 to 30 mm in length!), and our initial goal was to obtain small (nl quantities) of cytosol from the resting cell. To the surprise of our collaborating electrophysiologists, the cytosol could not be micropipetted. Later, when the muscle fibers were cut, no cytosol leaked out; instead, the cytosol behaved more like a gel than a sol. (Of course, this has been well known for over 30 years by neurophysiologists studying the giant squid axon; they routinely roll or squeeze out axoplasm gel like one squeezes toothpaste out of a tube, but the functional significance of this unexpected physical state of the ground substance in these cells has been largely ignored.) Many years ago, researchers such as Heilbrunn (1956) were fascinated by this behavior, but the techniques for studying the properties of the cytoplasm were not advanced enough to yield quantitative answers on its physical state *in vivo*. Today the situation is different. Using a variety of new and penetrating approaches, we now know that cytoplasm is remarkably viscous, in the range of 10 to 15 cP, not the 1 cP we normally assume, which is typical of water (Dix and Verkman, 1990). Similarly, under intracellular conditions, diffusion rates may be very different from those expected in simple solutions. Using an ingenious experimental approach, termed the "reference phase technique", workers in this area find that many proteins are diffusive, many are not (Paine, 1984), and that even small solutes (disaccharides are often advantageous in these studies) may show regional variation in diffusion rates (Mastro et al., 1984). Concentrations of small molecules and metabolites may be similar in different regions of the cell or may show systematic variations (Horowitz and Miller, 1984). Most important of all, all of the above, especially microviscosities in different regions of the cell, are not static physical states but vary with the activity or metabolic status of the cell (Paine, 1984; Dix and Verkman, 1990). Similar studies of structure, function, and dynamics of the mitochondrial matrix indicate an even more extreme situation: viscosity estimates are up to 40 times higher than for dilute solution, and these also change dramatically, depending upon the respiratory activity of the mitochondria. Highest viscosities are found during state 3 respiration, when the mitochondria are considered to be operating under substrate-saturating conditions and to have the "energized" or condensed ultrastructure (Scalettar et al., 1991). These observations raise the intriguing possibility that during rest $\rightarrow$ work shifts in skeletal muscles, transitions in the physical states of the cytoplasm and the mitochondrial matrix (analogous to phase transitions, gel to sol being an extreme example) could be instantaneously transmitted to all reaches of the cell. *If this were linked to O_2 supply to the cell (for example, through Ca^{++}), it would account for two of the chief experimental observations that frequently arise in this field: the direct relationship between O_2 availability and the ATP turnover rate and the near-instantaneous spread of the activation "signal" to essentially all protein components involved in integrated muscle function at once.*

These kinds of models for coupling O_2 sensing to regulation of ATP turnover obviously are still a long way from being complete or confirmed. Yet, in principle, they are attractive for several reasons: (i) they explain the frequently observed relationship between $\dot{V}O_2$ and perfusion, (ii) they supply a mechanistic basis for O_2 conformity (i.e., O_2 uptake rates varying with $[O_2]$ over broad concentration ranges usually very much higher than the K_m for O_2 for mitochondrial metabolism (Thurman et al., 1993)), (iii) they account for the coordinate regulation, upward or downward, of ATPases and ATP synthesis pathways with change in O_2 delivery, (iv) they supply a mechanistic explanation for up- or down-regulation of ATP turnover rate with minimal change in concentrations of intermediates in pathways of ATP utilization or of ATP production, and (v) they display interesting potential for strategic transfer to applied researchers, because these putative control systems seem to be highly conserved in evolution.

Although there is enough information to generate interesting models and thus to experimentally probe the mechanisms of O_2 regulation, the pathways and mechanisms for transducing the O_2 signal to both ATP demand and ATP supply pathways at once obviously are not currently known. What is known, however, is (i) that O_2 as a regulator seems to act external to the pathway of O_2 utilization in energy metabolism per se, in this way being similar to other exogenous regulators of ATP turnover and (ii) that the efficiency of such exogenous controls (i.e., the efficiency of energy coupling) is lower in anaerobically-driven muscles than in muscles relying on mitochondrial oxidative metabolism. Between-species comparisons (e.g., trout vs. tuna white muscles) as well as within-species comparisons (e.g., homologous muscles of power vs. endurance-trained athletes; mammalian cardiac muscle under hypoxic vs. normoxic conditions) indicate gradations of coupling efficiencies: the greater the dependence on aerobic metabolism, the greater the coupling efficiency; the more anaerobic the metabolism, the less efficient is the balancing of ATPases with ATP synthesis rates. At least a part of the basis for this gradation in coupling efficiency is to be found in the kind and amount of metabolic machinery that supports muscle ATPase functions. Muscles with high activities of specific kinds of glycolytic enzymes and of cytosolic CPKs depend upon pathways in which the molar amplification in ATP yield is modest (1 to 3 moles ATP per mole fuel mobilized) compared to O_2-based pathways in which molar amplification is much higher (e.g., 36 moles ATP per mole glucose oxidized). *A given imbalance in pathway fluxes relative to ATPase fluxes, therefore, is reflected in a much larger percent imbalance in energy coupling than is observed in aerobic pathways.* That, together with the absence of O_2 regulatory effects, is perhaps the most basic reason why the coupling of ATP demand and ATP supply in fast-twitch glycolytic muscle is less tightly regulated (or is less efficient) than in fast-twitch oxidative muscles or in slow-twitch muscles, both relatively rich in mitochondria. Since these systems-level differences are an expression of the protein and pathway compositions of different kinds of muscles, we are brought full circle—to the isoform basis of muscle structure and function, which is the last topic to which we will now turn.

8

Isoform Definition of Muscle Machines

INTRODUCTION

In the three major systems we have discussed to this point ((i) the muscle activation cascade and control system, (ii) the ATP utilizing contraction-relaxation cycle per se, and (iii) the ATP supply pathways supporting muscle work as such) the available evidence is consistent with mechanisms for minimizing diffusion dependence and clearly implies that natural selection is driving the system structure away from simple diffusional control of fluxes. Instead, the main flux determinants (whether they be ion fluxes, carbon fuel fluxes, or adenylate fluxes) over and over again appear to be the inherent efficiencies of interaction and exchange between the main protein components involved. Similarly, in all three above systems, the efficiencies of interaction and exchange clearly depend upon rather subtle (and at this time, rather poorly understood) structural constraints. That may well be why isoforms exist for essentially every protein-based step in the flow of information from motor neuron to muscle crossbridge cycling, from actomyosin ATPase to ATP-generating pathways, and from storage fuel depots to enzymes of catabolic pathways. Isoforms in essence define muscle types, for *specificity of muscle type means specificity of protein isoforms used in building the sarcomere, its activation control system, and its metabolic support pathways.*

DEFINING MUSCLE FIBER TYPES

This fundamental principle can be illustrated for any specific muscle type, for example, mammalian fast-twitch glycolytic muscle, which during differentiation is formed by the assembly of specific amounts and specific kinds of proteins. Minimally, these include:

1. Abundant plasmalemmal-type Ca^{++} channels in presynaptic terminals,
2. Very high densities of unique AChR (end-plate) channels, matched by a high ACh sensitivity and high AChE activity,
3. Very high densities of tissue-specific Na^+ channels, particularly near end plates and transverse tubules,
4. High and localized densities of tissue-specific delayed rectifier K^+ channels,
5. Very high densities of TT-type Ca^{++} channels with specific α-1 subunit ''binding'' sites for interacting with SR ''foot proteins,''
6. An expanded SR, with elevated densities of Ca^{++} release channel isoforms (''foot'' proteins) with specific sites for interacting with TT Ca^{++} channels and forming the allosteric EC coupling system, fast isoforms of calsequestrin, with bound tissue-specific isozymes of carbonic anhydrase,

7. High amounts of fast-type Ca^{++} ATPase in the SR, with Ca^{++} ATPase stoichiometry effectively less than 2 Ca^{++}/ATP hydrolyzed and with Ca^{++} pumps not subject to cyclic AMP-dependent protein kinase control,
8. High amounts of parvalbumin,
9. A specific tropomyosin isoform,
10. Specific troponin isoforms, especially troponin *c* which allows 2 Ca^{++} to be bound/per troponin,
11. A myosin isoform of very high ATPase activity, two ATP binding sites per myosin, plus a low K_m for ATP, with triplet electrophoretic pattern imposed by a unique combination of myosin light chain isoforms,
12. An actin isoform displaying high affinities for myosin, which strongly activates myosin ATPase activities,
13. High activities of muscle-type CPK isozymes (matching and coadapting with myosin ATPase catalytic capacity), with specific binding sites on the myofilaments (close to or on myosin ATPase), on the SR (close to Ca^{++} ATPase), and on the sarcolemma (close to Na^+ K^+ ATPase), but low activities of mitochondrial-specific CPK isozymes,
14. High enzyme capacities for taking up creatine, and high levels of GLUT4 from glucose transporter,
15. High activities of tissue-specific isozymes of glycogen phosphorylase and the tissue-specific isozyme cascade regulating the ratio of active and inactive form,
16. High activities of muscle isozymes of glycogen synthase and the enzyme cascade regulating its activities,
17. Relatively low activities of tissue-specific hexokinase,
18. High activities and tissue-specific distribution of PGM and PGI isoforms,
19. High activities of muscle PFK showing regulatory properties allowing for the flare-up activation kinetics required for rest—work transition, with binding sites on myofilaments,
20. High activities and tissue-specific distribution of aldolases with binding sites on myofilaments,
21. High activities and tissue-specific distribution of isomerases, GAPDHs, PGKs, mutases, and enolases, with binding sites on myofilaments and on band 3 proteins at least for phosphoglycerate kinase and GAPDH isozymes,
22. High activities and tissue-specific distribution of M and H type LDHs, M type subunits constituting over two-thirds of the total LDH activity,
23. Low activities of several related isozymes (alanine and aspartate aminotransferases, malic enzyme, PEP carboxykinase, FBPase) associated with either cytosolic-mitochondrial interactions or with resynthesis of glycogen from lactate,
24. High activities (relative to slow muscles) of cytosolic MDH and αGPDH isozymes,
25. Low activities of enzymes of the Krebs cycle and the ETS and of transporters and exchangers on the inner mitochondrial membrane, an ATP/ADP translocase anchored close to mitochondrial CPK binding sites, and finally
26. Low activities of a regulatory ATP synthase, showing positive cooperativity of ADP sites in catalysis.

This list for FG or Type IIb muscle is by no means complete. While covering only the minimal components, it already would require the regulated expression of over 100 genes. Thus, the list is long enough to drive home our point: when the right amounts of such right kinds of components are assembled, the resulting *muscle cells by definition are termed fast-twitch glycolytic or FG fibers.* Slow muscles are built up from completely different protein components (isoforms), involving activation of 100 or more different genes. Although the two muscle types are ultrastructurally similar, only actin is fully identical in both groups, and all these fundamental differences between fast and slow muscles appear to be conserved throughout the vertebrates. Similar isoform-based assemblies characterize different muscle types in cephalopods (Mommsen et al., 1981), in insects (Sacktor, 1976), in crustaceans (Ruegg, 1986), and presumably in all other animals complicated enough to express complex muscle-powered locomotory activities. A simple but powerful principle arising is that the *contractile (crossbridge cycling) capacities of any given muscle types, their Ca^{++} dependent activation and relaxation regulation systems, and the overall metabolic support systems underpinning such entire functional units are determined by the cell-line specific protein isoforms expressed.*

In this context, we can also better appreciate the plasticity of muscle structure and function. In principle, *the number of muscle types achievable might only be limited by the number of ways in which the above isoforms can be assembled.* If this indeed were the case, then the theoretical maximum number of fiber types expressed in any given muscle could become astronomically and thus unmanageably large. In practice, therefore, it is perhaps not surprising that *this maximum number is nowhere even approximated and probably never can be;* what is surprising is that just the reverse seems to be the case. The more finely tuned a muscle is, the more homogeneous it often seems to be, and this seems to be true both for very O_2 dependent muscles (some nearly obligatorily aerobic) and for very anaerobic ones, for muscles designed to generate power output or designed for other functions. This point is so fundamental that it is important to consider several kinds of such specialized muscle in some of detail.

NATURE'S FASTEST OXIDATIVE MUSCLES: ONE ISOFORM COMBINATION TO THE EXCLUSION OF ALL OTHERS?

Hummingbird flight muscle perhaps is the most highly O_2 dependent skeletal muscle found in the vertebrate world. As already mentioned, peak O_2 consumption rates can support ATP turnover rates of over 500 μmol ATP per gram per minute. These are the highest metabolic rates known in the vertebrates, and they can support a wing beat frequency approaching 80 Hz (J. Steeves, pers. comm.). This means that during hovering flight, muscle activation-contraction-relaxation cycles can be completed in less than 15 ms, and this astonishing frequency can be sustained for significant time periods apparently without any involvement of anaerobic metabolism. To achieve these enormous energy turnover and work rates, hummingbird flight muscles have exquisitely coadapted various key components of their biochemistry and physiology. Mitochondrial volume densities are up-regulated some five- to seven-fold compared to more usual oxidative muscles of endotherms; capillary abundance is similarly adjusted upward and is structurally modified (essentially wrapped around each muscle cell), so that O_2 delivery capacities are maximized to match the amplified capacities for mitochondrial O_2 uptake (Mathieu-Costello et al., 1992b). Fractional volume of sarcoplasmic reticulum is expanded by three- to four-fold compared to comparable muscles in other mammals and birds (Figure 8–1). From what we have learned above, this means a corresponding up-regulation of TT Ca^{++} channels, foot proteins, and Ca^{++} ATPases (all of which can also occur at higher densities). Numerous enzymes functioning in various pathways of oxidative metabolism (citrate synthase, HOAD, CPT, MDH, CPK, Hk, and PK, to mention a few) occur at higher

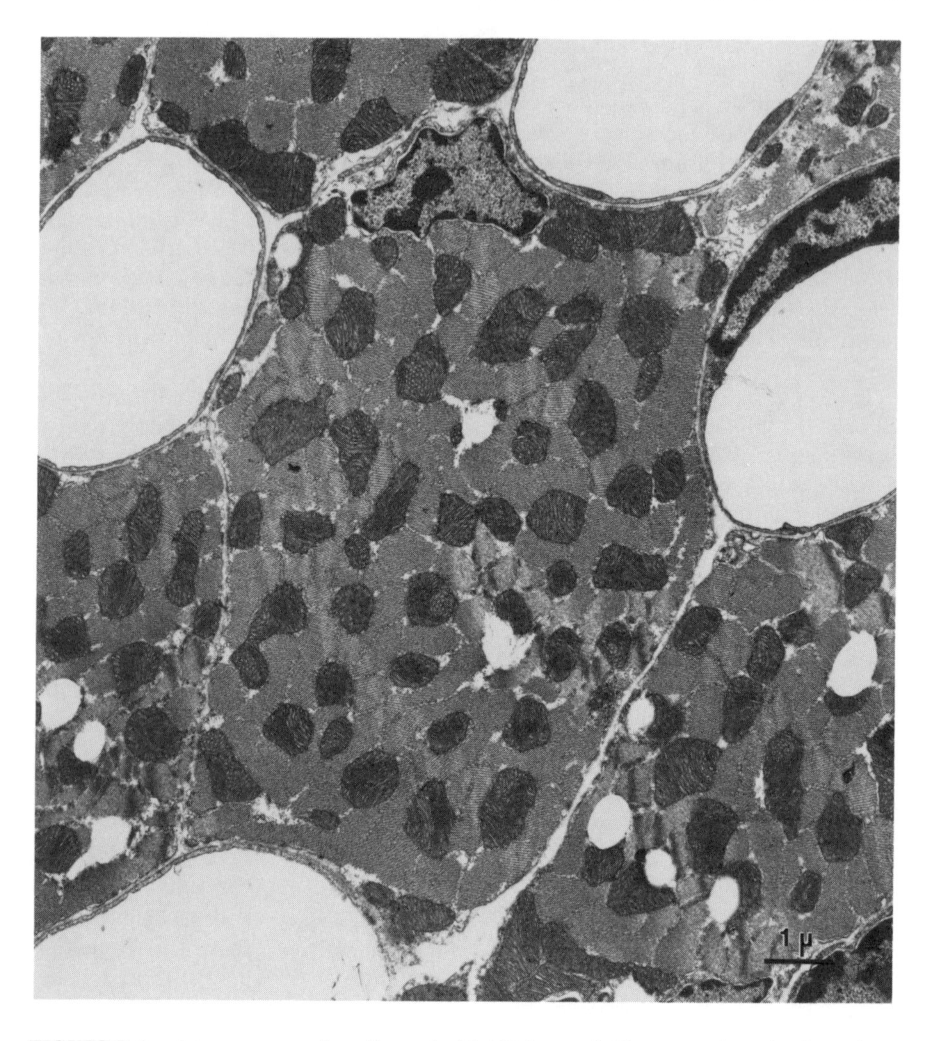

FIGURE 8–1. A transverse section of hummingbird flight muscle illustrates adaptation for perhaps the highest ATP turnover rates sustainable amongst vertebrates. To be able to do mechanical work at 80 Hz, the most striking requirement involves up-regulation of O_2 delivery capacities: high mitochondrial volume densities, higher than normal surface area of mitochondrial cristae, reduced diameter of muscle cells, and expanded capillarity (including the formation of a capillary manifold wrapping around the longitudinal axis of muscle fibers (Mathieu-Costello et al., 1992b). It is instructive that the SR amplification, while significant relative to other red-type muscles, is modest, especially when compared to muscles designed for high-frequency (sound-producing muscles) rather than for high mechanical work rates. Electron micrograph courtesy of Dr. Odile Mathieu-Costello.

activities per gram of muscle than found in any other vertebrate, cold- or warm-blooded! Lactate dehydrogenase activities, in contrast, are greatly down-regulated to one of the lowest values found in muscles of birds or mammals. Glycogen concentrations are seemingly in the normal range, but glucose availability is unusually high (probably because when flowers are available, hummingbirds preferentially feed on nectar). Intracellular fat is abundant and very adjustable according to season; premigratory birds store unusually large amounts of intracellular triglyceride in preparation for migrations that can be very long. Intracellular fat droplets often are seemingly preferentially deposited directly onto or immediately adjacent to mitochondria, their sites of catabolism.

In fact, in all ways that have been examined, hummingbird flight muscle seems to be honed for high performance (see Suarez et al., 1991 and references there in). The conclusion that arises from all available data is similar to themes already developed earlier in this book; namely, that the most finely tuned, metabolically active muscle in the vertebrate world is designed to optimize effectiveness of key interactions between protein components or links in the functional unit, i.e., to minimize diffusion distances between various key links or components in the functional unit. The conclusion seems valid for all stages in integrated function—for muscle activation, for allosteric excitation-contraction coupling (for TT Ca^{++} channel and Ca^{++} release channel interaction), for Ca^{++} troponin—tropomyosin activation, for muscle contraction per se, for Ca^{++} ATPase, for Na^+K^+ ATPase, for O_2 delivery, for carbon substrate delivery, for adenylate and utilizable phosphate bond cycling. The conclusion seems valid because otherwise we could not readily explain the high-speed performance of this muscle (less than 15 msec to complete each activation-contraction-relaxation cycle).

But what is the role of isoform-based fiber-type differentiation in this racing car version of vertebrate muscle? Interestingly, this is where the penny drops. As in other good avian flyers, hummingbird flight muscle is formed exclusively of one fiber type, fast-twitch oxidative (FOG) or type IIa fibers. Electron microscopy indicates remarkable cell-to-cell homogeneity, each muscle cell essentially identical to its neighbor. Although detailed isozyme studies have not been performed, at least one isozyme system has been examined—lactate dehydrogenase. Unlike the isozyme patterns often found in FOG fibers (dominated by M subunits), the LDH isozyme complex in hummingbird flight muscles has evolved toward one dominant band (the H4 isozyme), with a modest amount of a second band (the H3M1 band), indicating residual expression of the M gene in this tissue, exactly as in hummingbird heart (Hochachka et al., 1992). On the assumption that fiber ultrastructural homogeneity also implies homogeneity (or constant expression) of isoform components, we can tentatively conclude that in designing the most metabolically active muscle found among the vertebrates, evolution seems to have selected for a unique combination of component isoforms at the expense of all other possibilities. Instead of taking advantage of a seemingly enormous potential for making different kinds of muscle types, nature seems to opt for the exact opposite strategy.

Although space will not allow a detailed examination of the same situation in insect flight muscles, we should emphasize that the same conclusions arise for those species (locusts, bees, blowflies) whose flight muscles have been closely examined (see Casey et al., 1992). Indeed, since the metabolic rates here are even higher than in the case of hummingbird flight muscles, many of the above kinds of coadaptations of interacting components are pushed even further—again characteristically with muscles being honed to homogeneity. *Highest rates of Ca^{++} cycling, of adenylate cycling, and of O_2-CO_2 cycling seem to be achieved by restricting, not expanding, muscle cell heterogeneity.* At the cell level, because of the 100 or more genes involved in specifying a functional type IIa muscle cell, this means that the spectacular upregulation of performance is possible only by selective and extensive silencing of many other genes. Only one battery of genes and their regulators is turned on, that

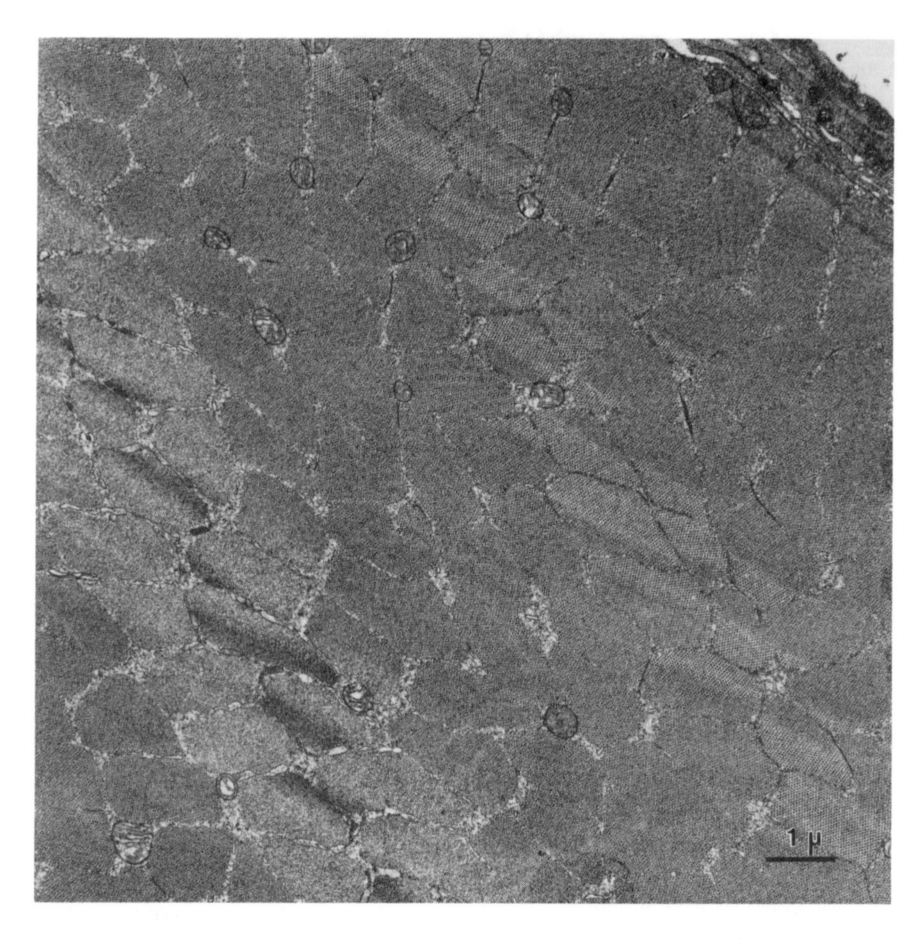

FIGURE 8–2. Transverse section of tuna white muscle as an example of an adaptation for high capacities for glycolytically and phosphagen-powered work rates. The main adaptations include enormous down-regulation of components involved in aerobic metabolism (mitochondria, capillaries), an increase in cell size, an up-regulation of myofilament volume densities, and an up-regulation of SR volume densities. Electron micrograph compliments of Dr. Odile Mathieu-Costello.

specifying the unique isoform components that constitute a fast-twitch glycolytic-oxidative type of muscle cell. Whether other fiber type-specific batteries of genes can be turned on under appropriate conditions in this system, as in other species (Pette and Staron, 1993; Booth and Tseng, 1993) remains an unanswered question.

FISH WHITE MUSCLE: AN ANAEROBIC TYPE OF MUSCLE DISPLAYING EXCEPTIONAL COMPOSITIONAL HOMOGENEITY

Similar conclusions arise from analysis of fast-twitch glycolytic muscles evolved for short-term explosive work. Best examples come from fishes where these fibers are spatially differentiated from slow oxidative fibers (the red muscles of the fish literature). In some species, such as the fast start-and-sprint champions of the piscine world, such as the pike, the entire swimming musculature is a uniformly white or fast-twitch glycolytic system, with red (or slow-twitch) muscle reduced to vanishing levels (see Moyes et al., 1992 and literature therein). Again, ultrastructural examination of fish white muscle shows that each cell is essentially identical to its neighbor

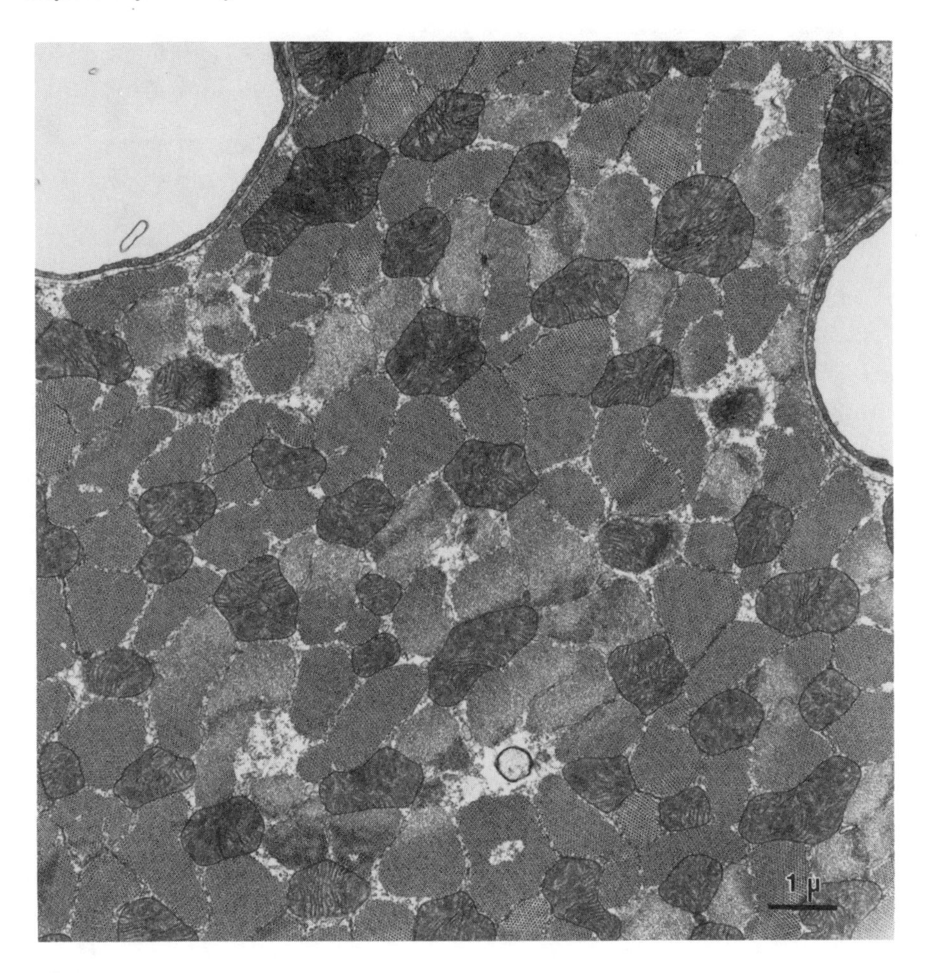

FIGURE 8–3. Transverse section of tuna red muscle as an example of an adapation for sustained high work capacities. The contrasts with the anaerobically powered white muscle shown in Figure 8–2 could not be more striking. The adaptations occurring include explosive up-regulation of mitochondrial volume densities, smaller cell size, abundant SR, and expanded capillarity. As in hummingbird flight muscle, the capillaries also form a manifold surrounding the longitudinal axis, two fibers in each manifold, so as to increase surface area and the time available for gas exchange (Mathieu-Costello et al., 1992). Electron micrograph courtesy of Dr. Odile Mathieu-Costello.

(Figures 8–2 and 8–3), and on the reasonable assumption that ultrastructural homogeneity implies a constant expression of all other components of the system, it is a fair conclusion that in designing one of the most explosive of burst muscle machines in nature, evolution in each species has again selected one or only a few unique isoform combination, out of the astronomically large number of statistically realistic possibilities.

SOUND-PRODUCING MUSCLES: STRUCTURALLY HOMOGENEOUS MUSCLES DESIGNED FOR HIGH FREQUENCY, NOT POWER OUTPUT

Another example of muscles that have evolved to a state of very high perfection are the sound-producing muscles of insects, fishes, birds, and bats. Some of these are

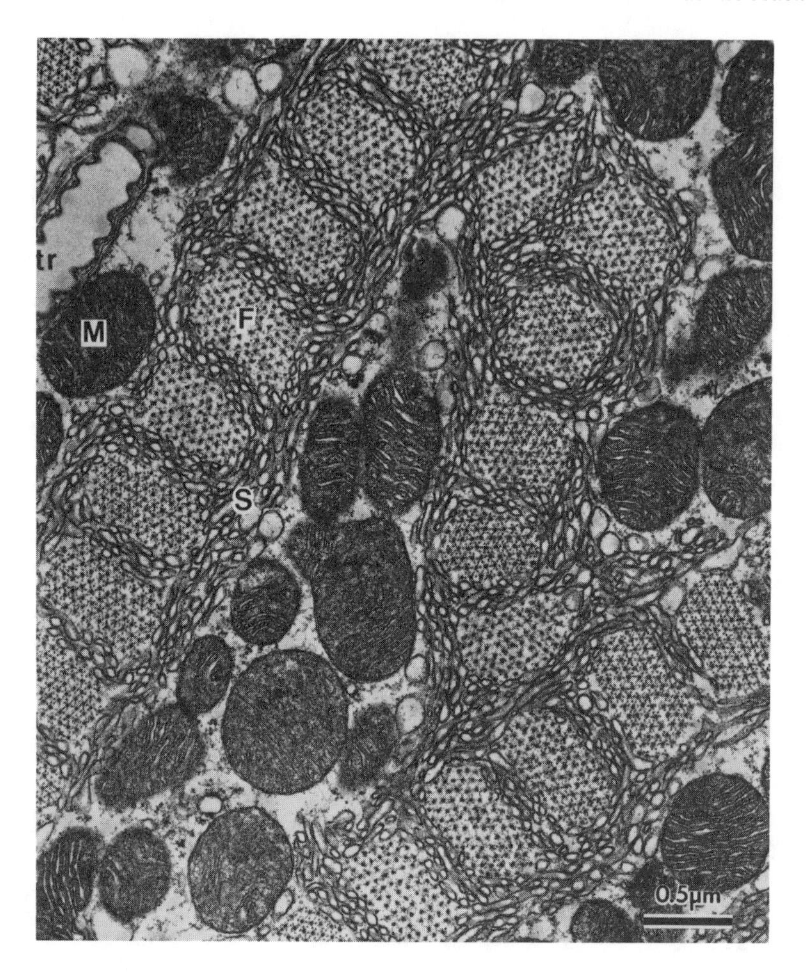

FIGURE 8–4. Tranverse section of tymbal (sound-producing) muscle of the cicada as an example of an adaptation for high-frequency performance. The capacity for 550 Hz operation comes with the price of immense down-regulation of myofilament volume densities (this is only possible because moving the tympanic membrane to generate sound is not very demanding work!). The most striking additional adaptations are up-regulation of mitochondrial and SR volume densities, each occupying approximately one-third of the cell volume. Scale bar is 0.5 µm. Modified from Josephson and Young (1985). Electron micrograph courtesy of Dr. R. K. Josephson.

based on an oxidative metabolism; others are based on a strongly anaerobic metabolism (see Hochachka et al., 1988). In all cases, however, what is observed is a remarkable homogeneity of composition and ultrastructure. The extreme case is probably the sound-producing muscle of the cicada. Able to operate at up to 550 Hz, this muscle is considered to have gone about as far as biochemically possible in achieving its remarkable properties. The limiting constraint seems to arise from the balance required between ATP synthesis capacities (mitochondrial volume densities) and Ca^{++} cycling capacities (SR volume densities). Further up-regulation of either component seems unlikely without down-regulation of the partner in the ATP demand—ATP supply cycle (Hochachka et al., 1988). Be that as it may, what is perhaps most striking about this muscle is that it is extremely homogeneous, formed of a single

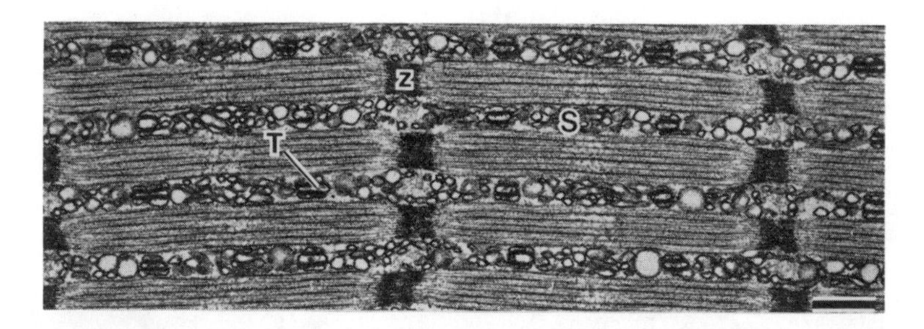

FIGURE 8–5. Longitudinal section of tymbal muscle of the cicada showing the expanded SR and its relationship to the myofilaments and the transverse tubules (T). The transverse tubules interact with the SR at repeat positions that are one-fourth of the distance between Z-lines. The precision of this interaction is striking; it implies an underlying homogeneity of isoform components and undoubtedly contributes to the efficiency of Ca^{++} pulsing between the SR lumen and the cytosol. Scale bar is 0.5 μm. Modified from Josephson and Young (1985). Electron micrograph courtesy of Dr. R. K. Josephson.

cell type (Josephson and Young, 1985). Again, instead of taking advantage of the enormous potential for numerous isoform-based cell types, nature seems to opt for a unique combination, to the exclusion of all others (Figures 8–4 and 8–5).

RATTLER MUSCLE OF THE RATTLESNAKE

A third example of muscles that have evolved toward a high state of perfection are the rattler muscles of rattlesnakes. Designed for high frequency but low power output performance, these muscles display up-regulation of capillary volume density and are jam-packed with mitochondria, approaching densities characteristic of heart rather than skeletal muscles (Figures 8–6 and 8–7). Because of an unusual combination of requirements—very high rattling frequency sustained by mitochondrial metabolism—the high mitochondrial volume density is matched by an even more striking up-regulation of SR. In most mitochondria-rich slow muscles, SR constitutes about 2 to 4% of the cell volume. In contrast, the SR of the mitochondria-rich, high-frequency rattler muscle constitutes up to 30% (!) of the cell volume, creating a quite unusual ultrastructural design (Conely, K. and S. Linstedt, pers. comm.). As in the examples given above, the muscle cells of the rattler display rigid uniformity, each cell a mirror image of its neighbor, and the same principle as before is evident: in an aerobic muscle selected for very high frequency rattling rather than for high mechanical power output, the defense properties required are attained not by expanding muscle cell heterogeneity, but by restricting it.

BRAIN HEATER ORGAN: A STRUCTURALLY HOMOGENEOUS MUSCLE DESIGNED FOR THERMOGENESIS, NOT POWER OUTPUT

A relatively rare but highly interesting kind of muscle has been adapted for the unusual function of thermogenesis in a number of large, fast-swimming fishes (mainly the marlin and swordfishes). Designed for using the energy released during high rates of ATP cycling for generating heat and warming the brain (rather than for power output), this muscle is packed with SR and mitochondria (see Block, 1991; Block et al., 1993). The SR is rich in Ca^{++} ATPase, which despite its high catalytic capacity, is unable to keep up with a specially adapted (very leaky) Ca^{++} release channel,

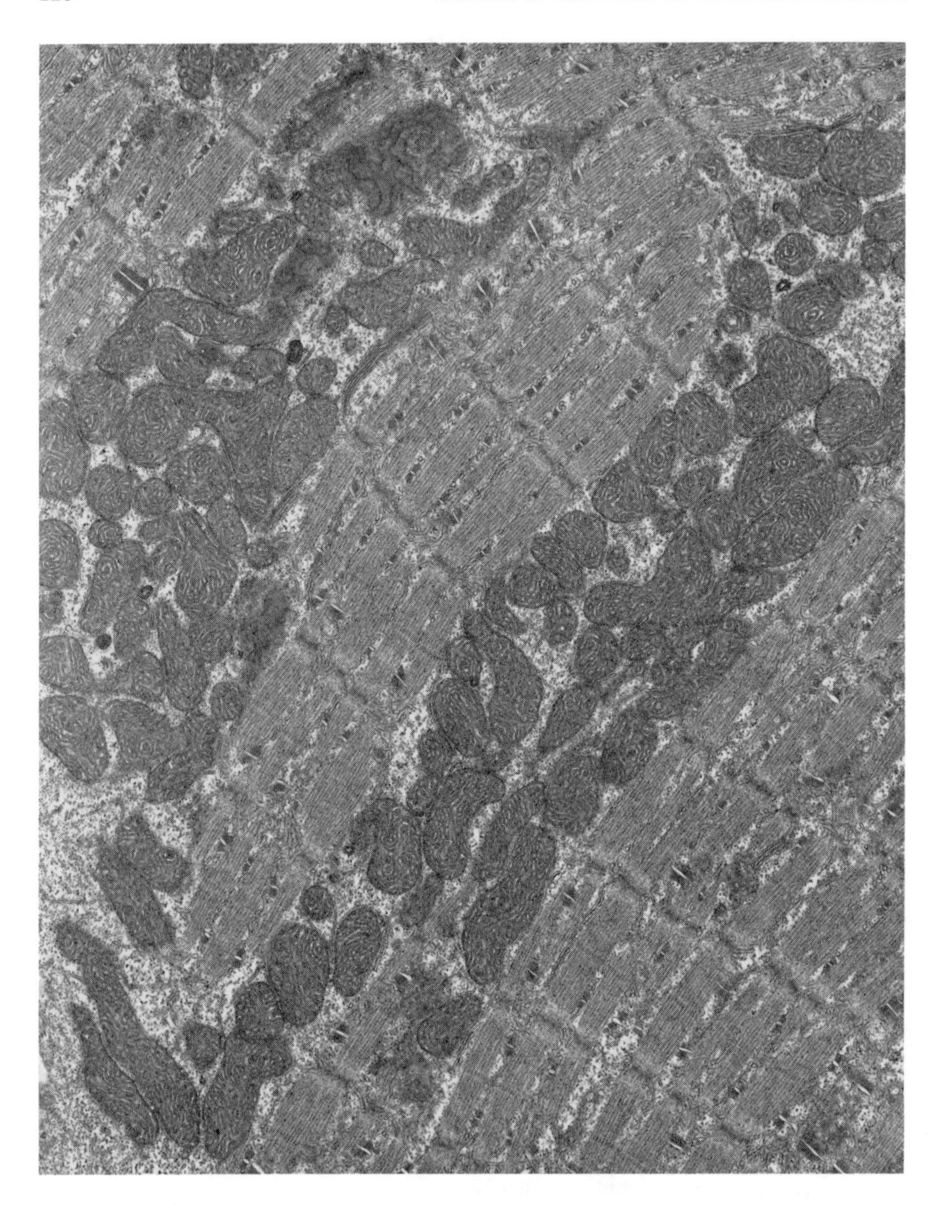

FIGURE 8–6. Longitudinal section of rattle muscle of the rattlesnake as a vertebrate example of a muscle adaptation for sustained high-frequency, but low power output function. Main adaptations include up-regulation of mitochondrial volume densities, which are about equivalent to the volume densities of the myofilaments. Electron micrograph courtesy of Dr. S.L. Lindstedt, P.J. Schaeffer, and M.L. Sellers.

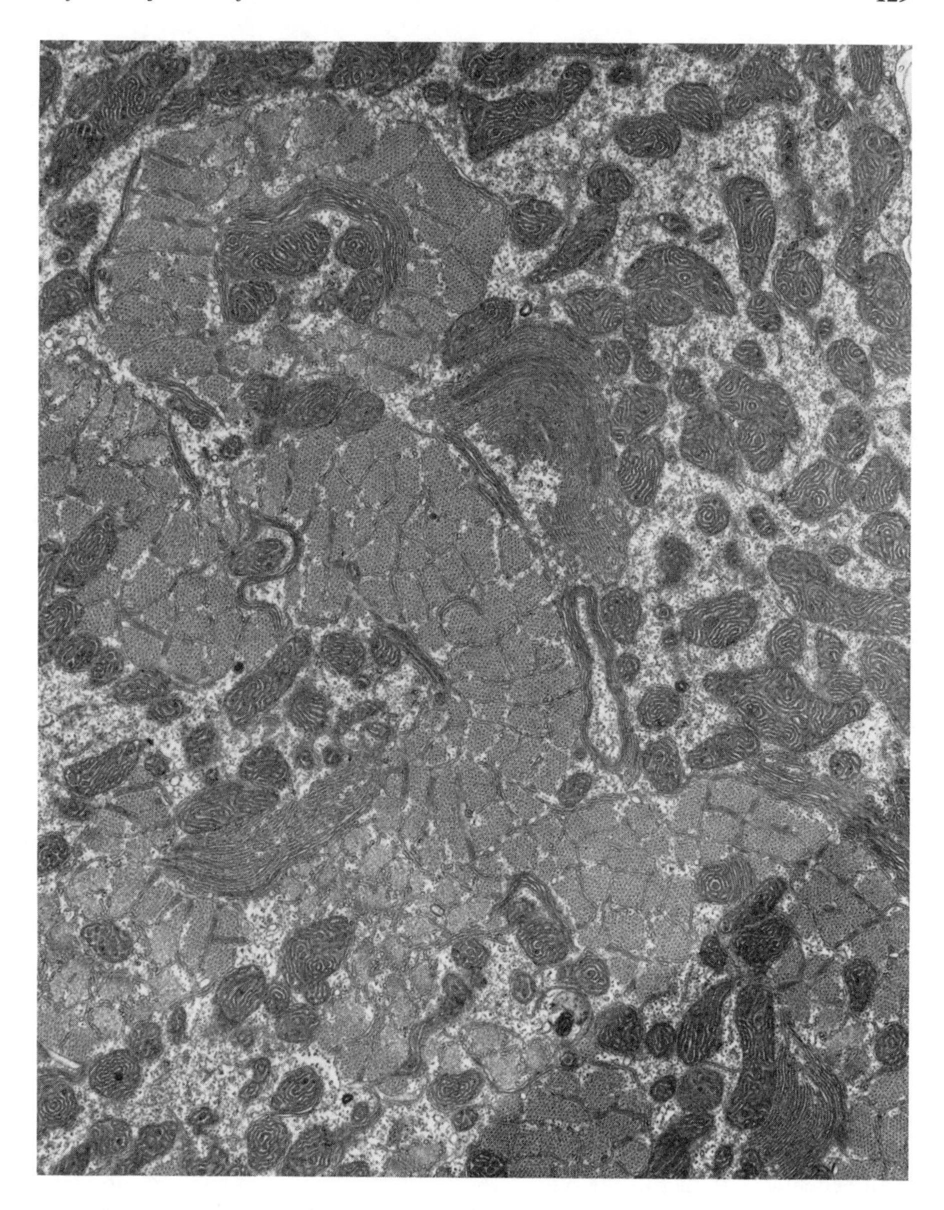

FIGURE 8–7. Transverse section of rattle muscle of the rattlesnake as a vertebrate example of a muscle adaptation for sustained high-frequency, but low power output function. The same high volume density of mitochondria is observed as in Figure 8–6; in this section, however, the extensive SR surrounding the myofilaments is illustrated as is the expansion of the mitochondrial cristae. Electron micrograph courtesy of Drs. S.L. Lindstedt, P.J. Schaeffer, and M.L. Sellers.

while the mitochondria are rich in the enzymes of oxidative metabolism. ATP synthesized in the latter is dissipated through Ca^{++} ATPase and the semifutile cycling of Ca^{++} between the cytosol and the SR. As in the examples given above, what is striking about the ultrastructure and composition of this modified muscle is that it is again notably homogeneous. Each cell is essentially a carbon copy of its neighbor. In the design of this muscle, evolution is again selecting for a unique isoform combination for the basic functional unit—isoform based cell-to-cell heterogeneity being rejected.

ELECTROPLAX: A STRUCTURALLY HOMOGENEOUS KIND OF MUSCLE DESIGNED FOR ELECTRICAL DISCHARGE, NOT POWER OUTPUT

A final example that comes to mind illustrating our theme is the modified muscle in electric fish that forms the electric organ. Perhaps more so than all the other examples we have chosen, the design of the electroplax expresses cell-to-cell ultrastructural and compositional homogeneity. Unlike the examples above, where the end product of the muscles considered includes work (high power output), sound (high frequency vibrations), or heat (high rates of ATP turnover dissipated in Ca^{++} cycling), in the electroplax, the end product of this specialized muscle machine is electric discharge (see Blum et al., 1991). For this specialization compared to any of the above, clearly quite different protein isoforms had to be selected (ion-specific channels, ion-specific pumps, and other membrane-based proteins especially had to be kinetically and probably structurally tuned up; many if not all of the genes for muscle contractile machinery per se presumably were silenced). The coupling of ATP-generating pathways (such as CPK) to the dominant ATPases (the Na^{+} K^{+} pumps), especially the tightness of interaction between these kinds of protein components, seems to be particularly efficient (Blum et al., 1991). Whatever the isoform composition of each link in the overall functional electroplax unit, it has been clear for years that cell-to-cell homogeneity is the rule; isoform-based cell-to-cell heterogeneity again is either totally excluded or at least seems to be extremely rare.

WHY MUSCLES SPECIALIZE INTO SO FEW DIFFERENT FIBER TYPES

If the above analysis is correct, it firmly supports the conclusion that the end point of evolution of muscles for the production of power, sound, heat, or electricity selects not for enormous isoform-based cell-to-cell heterogeneity, but rather for the opposite; a unique and singular cell type formed presumably of a unique assortment of protein isoforms. This extreme end point, however, is rarely reached in skeletal muscles of most vertebrates. Rather than a rule of cell-to-cell similarity and homogeneity, two muscle types are almost universally observed in vertebrates, although additional types, intermediate to the fast and slow extremes defined above, are probably also expressed. We suggest that the isoform design of the overall system is one reason why the realized number of muscle types is only a minute fraction of the maximum number theoretically possible. Just as the drive shaft of a sports car would not do in a cement truck, troponin *c* isoforms in fast muscles may not be suitable for slow muscles; fast muscle Ca^{++} ATPase may be debilitating to slow muscles, while slow muscle presynaptic Ca^{++} channels would simply not work well enough in fast muscle, and so forth. Point mutation-based muscle diseases often impact on only one link in integrated muscle function, leading to one isoform alteration in the overall system structure, yet they can have devastating effects. An increased activity of a Ca^{++} leak channel in muscle cells in Duchenne dystrophy supplies an interesting insight into this very problem. In many of its properties, this Ca^{++} release channel is similar to the usual isoform found in other muscle cells; however, the open probability of this release channel is markedly increased in dystrophic cells. This defect is considered to be the most likely molecular explanation for the elevated intracellular Ca^{++} concentrations in human dystrophic myotubes (Fong et al., 1990). By the same token, probably only modest modifications in the SR Ca^{++} release channel of the heater organs of tuna and marlin are required to convert a highly regulated Ca^{++} cycling between this channel and SR Ca^{++} ATPases to a relatively uncontrolled Ca^{++} efflux from SR and a kind of energy short circuit that is the molecular representation of this particular biological furnace. Similarly, simple mutations affecting the isoform composition of metabolic pathways (on the ATP supply side of the system) are well known to have extremely disruptive effects. The literature here is large, and this is not the place to review it. Suffice it to mention an LDH myopathy—lack of the M-type LDH subunit in skeletal muscle—described by Kanno et al., (1980) and now known to be caused by a 20 base sequence deletion in exon 6 of the M4 LDH gene. Individuals lacking M4 LDH in skeletal muscle are found to display abnormally low capacity for anaerobic muscle work because of abnormally low capacity for anaerobic glycolysis and lactate production. Interestingly, the effective down-regulation of the glycolytic pathway in muscles of M4 LDH deficient subjects leads to a tighter coupling between ATP demand and ATP supply during muscle work; as a result, change in work rate of skeletal muscles is accompanied by lesser changes in the concentrations of phosphate intermediates, such as PCr, P_i, and the adenylates, than are found in muscles with more vigorous glycolytic capacity (Hochachka, P.W., G. Zhu, C. Stanely, and T. Kanno, unpublished data).

More recently, molecular biologists have been able to tinker with the various parts of macromolecular protein components. One approach is to modify structural components of muscle using a variety of physiological stimuli that preferentially activate or silence specific muscle genes (Booth and Tseng, 1993). A more direct approach is to directly alter gene products. One such study involved putting together chimeric Ca^{++} ATPases—parts from the fast muscle isoform, parts from the slow muscle isoform. While the reason MacLennan and his coworkers (1990) launched this study is to better understand and probe structure–function relationships in the oligomeric holoenzyme, their work simultaneously exposes the principle that we are reviewing here, namely, that nature must take exquisite care in the way it assembles the macromolecules that run skeletal muscle because not all combinations are equally effective. In this particular case, not all combinations of different Ca^{++} ATPase parts are equally effective as Ca^{++} ATPases in different (fast vs. slow) types of muscles. Similarly, even if nature had at least 3 Ca^{++} release channels from which to choose in designing muscle cells, typically the fastest muscles especially in lower vertebrates express only one isoform, the α-isoform (O'Brien et al., 1993), which can physically interact with the transverse tubule Ca^{++} channels (allosteric mediation of EC coupling).

Isoforms apparently cannot be simply randomly assembled to form an efficient, workable contractile unit. Certain combinations work best for certain jobs; other combinations work for other jobs. This interesting situation means that *isoforms are both the means by which muscle plasticity is achieved and the means by which plasticity is constrained.*

It is intuitively easy to understand these kinds of isoform-based constraints on the way muscles can be assembled. Plasticity, however, is another matter. The role of isoforms in extending muscle adaptive possibilities is not intuitively obvious and requires detailed knowledge of the functional properties of isoforms on both the demand and supply side of bioenergetics. One way to explore this issue is to review a specific example and see if any general principles can be deduced in the process. A case in point bringing some of these issues into focus is the lactate paradox—reduced glycolytic activity during exercise in chronic hypobaric hypoxia. Although the term implies a narrow phenomenon, the lactate paradox is actually an expression of molecular and metabolic adaptations for exercise under O_2 limiting conditions that maximize the amount of work achievable on a given amount of O_2 and other fuels. Analysis of how this is achieved simultaneously exposes the role of isoforms in determining the plasticity of muscle function. To put this in proper context, we need to begin with the empirical observations—lactate production during exercise under normoxia vs. hypoxia—which explain how and why workers in this field defined the paradox in the first place.

GLYCOLYTIC FUNCTION IN CHRONIC HYPOBARIC HYPOXIA

During incremental exercise to fatigue in normal lowland subjects, plasma and working-muscle levels of lactate begin to increase significantly at work rates of perhaps 60% of maximum, and at fatigue, plasma levels can exceed 15 m*M*. If the same subjects are tested under acute hypoxia for any given work level, they form and accumulate more lactate, which can be viewed as a special version of the Pasteur effect. However, after a period of acclimation, the [lactate] pattern is shifted back toward that found under normoxic (sea level) conditions. This acclimation effect is even more pronounced in long-term adapted altitude natives. For example, in their normal high altitude environment, Quechua Indians of the high Andes accumulate plasma lactate to concentrations that are one-third to two-thirds the values observed in lowlanders under otherwise identical exercise conditions (see Hochachka et al., 1991; Matheson et al., 1991). This phenomenon of low lactate accumulation despite hypobaric hypoxia, termed the lactate paradox by workers in this field (West, 1986; Hochachka, 1988c), was first discovered over 50 years ago, and in Quechuas is known to be unaffected by acute exposure to hypoxia or by acclimatization. NMR and metabolic biochemistry studies suggest that closer energy demand—energy supply coupling in Quechuas allows change in work rate with modest change in adenylate concentrations and thus minimal activation of glycolytic flux to pyruvate. The basis for the better energy coupling appears to be more effective integration (i) between glycolysis with mitochondrial metabolism and (ii) between ATP utilization and ATP production (i.e., between mitochondrial ATP production and ATP utilization by Ca^{++} ATPase and actomyosin ATPase isoforms (Hochachka et al., 1991; Matheson et al., 1991).

ROLE OF MDH AND LDH IN CHRONIC HYPOBARIC HYPOXIA

Although these data help to understand how Quechuas can either work harder or work longer under hypoxic conditions, by themselves they do not provide a mechanism accounting for reduced pyruvate to lactate flux per se. A follow up study (Hochachka et al., 1992) focused on enzyme activities in muscles from Quechuas because the lactate paradox in principle can be explained either (i) by limiting pyruvate production or (ii) by limiting LDH competitiveness for NADH, thus assuring a low pyruvate flux towards lactate. One way of evaluating the first alternative is by comparing the catalytic capacities for pyruvate production (PK activity) to the catalytic capacity for pyruvate reduction to lactate (LDH activity). Since PK/CS or PK/HOAD ratios are in the normal range for human subjects, we can conclude that the capacity for PK-catalyzed production of pyruvate in Quechuas has not been seriously compromised. Yet the activity ratios of PK/LDH in Quechua muscles are inordinately high by skeletal and even cardiac muscle standards due to low LDH activities. To put this into perspective, we should emphasize that in fast-twitch muscles, the PK/LDH ratio ranges from about 0.2 to 0.5; in slow oxidative muscles, the PK/LDH ratio is in the range of 0.8 to 1.0. Although these correlations are widely represented, it is important to point out that they are based on *metabolic,* not cell type, constraints and can be adapted when the need arises. For example, Weddell seal heart, a tissue unusually well adapted for surviving periodic O_2 limitation, displays a PK/LDH activity ratio of about 0.2, essentially equivalent to pure white muscle (see Hochachka, 1980) In this context, it is instructive that the muscle PK/LDH ratio in Quechua natives is higher than all of these values and in fact is higher than any values we are aware of from vertebrate muscles, except for hummingbird flight muscles. The latter are very aerobic muscles dependent on fluxes of glucose $\rightarrow$ pyruvate $\rightarrow$ CO_2 + H_2O for sustaining very high aerobic metabolic rates. In hummingbird flight muscle, the capacity to generate pyruvate (the PK activity) exceeds the capacity to convert it to lactate (the LDH activity) by a full three-fold.

Although we are getting closer, these data again by themselves cannot fully account for the lactate paradox, for *in vivo* LDH never functions under pyruvate saturation; what, therefore, is to prevent LDH function at higher [pyruvate] and thus "normal" lactate production in these muscles? The answer seems to lie in NADH limitation, alternative (ii) above. This is because LDH down-regulation in Quechua muscle correlates with maintenance of high total MDH activity per gram muscle. Under aerobic conditions, cytosolic and mitochondrial MDH isoforms serve in the shuttling of coenzyme between the cytosol and the mitochondria via the so-called malate-aspartate shuttle. High capacity for this function is important because high MDH activities can outcompete LDH for NADH and thus supply what we take to be the crucial "brake" upon rates of lactate formation. The impact of this kind of competition would be particularly large under aerobic conditions when NADH concentrations are considered to be low and potentially limiting. Interestingly, up-regulation of MDH activities (with simultaneous down-regulation of M-type LDH activities) occur in human leg muscles during endurance training (Burke et al., 1977; Chi et al., 1983) with a simultaneous reduction in lactate formation for any given intensity of exercise (Hurley et al., 1984). They are even further adjusted (the long-term adaptation) in hummingbird flight muscle, where MDH/LDH ratios reach values as high as 15 compared to about 2 for Quechua muscles.

ROLE OF LDH ISOZYMES IN CHRONIC HYPOBARIC HYPOXIA

In heart and in some slow oxidative muscles, low rates of lactate production are also assured by LDHs that are better lactate oxidases than they are pyruvate reductases. These can be identified by pyruvate-inhibition of heart-type LDH isozymes. This characteristic is also expressed by LDHs in Quechua and in hummingbird muscles (Hochachka et al., 1992), and can be viewed as a secondary or backup kinetic mechanism for minimizing lactate production during muscle work in both systems. The flip side of this argument arises from the data on Weddell seal heart; here the proportion of M-type subunits is up-regulated, and the LDH reaction as a result is less sensitive to pyruvate inhibition (adaptive whenever the tissue has to rely upon anaerobic glycolysis).

These enzyme data parallel known trends in lactate metabolism: fast-twitch muscles typically can generate large amounts of lactate rapidly; slow-twitch muscles generate lactate slowly or utilize lactate, depending on conditions (Connett et al., 1985; Brooks, 1986); cardiac muscles typically utilize lactate, generating it only under O_2 limiting conditions (Drake, 1985). That is why we conclude that this enzyme organization is a key contributor to the lactate paradox; low [lactate] is observed primarily because regulatory controls on isozyme kind and isozyme amount set limits on pyruvate reduction to lactate.

Taken together, the data on relative LDH catalytic capacity and on the pyruvate sensitivity of the reaction indicate that skeletal muscle in the Quechua is defended against over-production of lactate in the same way as the heart is defended in most mammals studied to date: by maintaining (with specific isozymes of PK, MDH, and LDH) high PK/LDH and MDH/LDH ratios. High PK/LDH ratios assure mitochondrial metabolism of a ''pyruvate push'' from glycolysis, while high MDH/LDH ratios assure that most cytosolic NADH is oxidized by MDH (for transfer to the electron transfer system (ETS)), not by LDH, which would lead to lactate accumulation. As in the heart, rates of ATP utilization are very precisely balanced with (aerobic) ATP production, so that large-scale changes in work can be sustained with less perturbation in the adenylates and phosphagen concentrations than occurs in lowlanders. Since changes in the adenylate levels under these conditions serve to activate glycolysis, the implications are that in Quechua muscle glycolytic activation is not as extreme as in lowlanders, and hence a better balance between glycolytic and Krebs cycle flux is maintained at any given work rate.

THE FIXED NATURE OF THE LACTATE PARADOX IN ANDEAN NATIVES

While short-term (training or acclimatization) adjustments in enzyme levels or fiber type composition can occur, these have never been shown to be large enough to account for the drastic differences in PK/LDH ratios observed in altitude-adapted natives. Thus, we would predict that if these enzyme activities are the basis for the lactate paradox, then the paradox would not deacclimatize over time periods of days to weeks at sea level (previously confirmed by Hochachka et al., (1991)) and indeed probably would be stable over generations (i.e., genetically determined, as observed for many other enzyme-based properties). New data (Hochachka et al., 1992) show that this indeed is probably the case. Comparisons of Quechuas at altitude (hypobaric hypoxia) and at sea level (normobaric conditions) indicate similar lactate production and accumulation for a given work load, unlike the situation of lowlanders when compared under hypoxic vs. normoxic conditions (Hochachka P.W. and C. Stanley, unpubl. data).

ISOFORM BASIS FOR PLASTICITY OF MUSCLE FUNCTION

To sum it all up, then, it appears that the present day representatives of the Incas display a muscle metabolic organization at the pyruvate branchpoint that is fine-tuned toward a pattern found expressed in extreme form in hummingbird flight muscle. In today's Incas, despite chronic hypobaric hypoxia, the capacity to generate lactate is down-regulated in order to maximize the yield of ATP/mole of fuel burned and avoid the undesirable end-product aspects of anaerobic metabolism. In hummingbird flight muscles, the capacity to generate lactate is down-regulated in order to sustain super-high ATP turnover rates while also avoiding the undesirable end-product effects of anaerobic glycolysis. These adaptive mechanisms may be operative in endurance-trained athletes, since in many ways their muscle metabolic organization is moved toward that in Quechuas but of course usually does not and presumably cannot reach the same end points. The parallelism observed in these different settings, however, beautifully illustrates the fundamental principle that much skeletal muscle plasticity (which allows adaptation to reach these end points in the first place) is based upon adjustments in isoform composition and content. For the examples chosen above, selection apparently is acting in at least six locations in the energy demand-energy supply cycle of skeletal muscle, in each case, both the isoform patterns expressed and the total catalytic activity (isoform content) being crucial to the adaptation. Minimally, the adaptation selects for

1. Specific kinds and amounts of myosin ATPases,
2. Specific kinds and amounts of Ca^{++} ATPases,
3. Specific kinds and amounts of pyruvate kinases,
4. Specific kinds and amounts of lactate dehydrogenases,
5. Specific kinds and amounts of cytosolic malate dehydrogenases, and
6. Specific kinds and amounts of mitochondrial malate dehydrogenases.

Additional isoform loci (such as the aminotransferases required for the effective transfer of reducing equivalents from cytosol to the mitochondria) may well be subject to coordinated regulation, but to this point have not been properly analyzed. More to the point, these specific examples illustrate that isoforms not only constrain but also amplify adaptive opportunities. Once again, we see why isoforms are a means by which muscle plasticity is achieved and a means by which it is constrained.

Many workers in this field would argue for a different emphasis in reviewing these kinds of adaptational data. They would point out that a first adaptational line involves change in the proportion of different fiber types in the muscle bed under consideration; they would then argue that change at such a cell (or fiber type) level automatically assures the metabolic adjustments we are discussing. Whereas this interpretation is acceptable as far as it goes, it clearly is inadequate to account for all the data. For example, the percent of slow-twitch fibers in vastus lateralis of Quechuas is not very different from that in endurance-trained lowland athletes; however, the PK/LDH ratio is higher than ever before found for any other humans, indeed higher than for many other vertebrate muscles! Similarly, an exhaustive analysis of some 13 enzyme profiles in over 20 muscles of diving mammals (seals and whales) indicates many profound adaptations that occur independent of fiber type distribution (Hochachka and Foreman, 1993). For such reasons, we feel that the older view of muscle plasticity needs some revision: a first line of adaptation may well involve changing percent of specific fiber types as a kind of coarse-level adjustment. The true plasticity of muscle is then expressed by second-level adjustments of isoform kinds and isoform concentrations, which serve to fine-tune the system for its particular environmental conditions.

THE ISSUE OF EMERGENT PROPERTIES

Biologists—especially those with an integrative approach—will recognize an almost time-honored theme that has been recurring throughout most of this text: the theme of emergent properties. While we have not been making too much of an issue of this idea, it nevertheless is evident from our analysis that *the functional characteristics expressed by interacting components at practically all levels of integrated muscle function are emergent properties in the sense that they cannot be expressed (often are basically undetectable) when each component is analyzed in isolation from its normally-linked neighbors.* This principle holds (i) at the level of design and assembly of unique isoforms of channels, transporters, exchangers, pumps, metabolic enzymes, and metabolic enzyme complexes (ii) at the level of hormone and neurotransmitter receptors, (iii) at the level of signal transduction from the sarcolemma, the transverse tubules, and the SR Ca^{++} release channel, (iv) at the level of information transfer (via the troponin-tropomyosin complex) to the contractile elements, (iv) at the level of the contractile machinery per se, (v) at the level of high energy phosphate transmission from ATP synthesis sites to ATP utilization sites (CPK isoforms interacting with specific cell ATPases), and finally, (vi) at various levels in the design of multiple interacting metabolic pathways designed to integrate ATP supply with ATP demand. While these emergent properties of integrated systems cannot be exposed, analyzed, or fully explained within purely reductionist experimental approaches (one cannot study in isolation what is not there), the reductionist attack, nonetheless, often adds enormous power and perception to defining the dimensions, boundaries, and mechanisms of the system structure and function. One need but recall the elegant studies of Ca^{++} channel isoforms in fast-twitch muscles compared to cardiac muscles; however, even in this case, knowledge of the integrated characteristics of EC coupling in these two different kinds of muscles necessarily preceded and then laid the foundation for the subsequent reductionist analyses with chimeric channels. The full power of the reductionist approach is realized only when it is reincorporated into the integrationist one. Emergent properties of linked and interacting components are exposed and can be analyzed and fully explained only within integrationist paradigms.

The analogy with working machines seems entirely appropriate. It is biological machinery we are talking about, but machinery, nevertheless. As in any man-made counterpart, fine-tuning (of isoform content and composition) is of course possible and may be desirable, but large-scale change in any one component of the overall machine system may well be expected to reverberate throughout the whole system. That is why the effects of any one of a host of modest mutations (causing single but large magnitude change in any one component of the system) are, in machinery analogy, like a spanner in the works.

Misplaced spanners are intolerable in man-made and in muscle machines.

REFERENCES

Almers, W., P.R. Stanfield, and W. Stuhmer (1983) Lateral distribution of Na and K channels in frog skeletal muscles. Measurements with a patch-clamp technique. *J. Physiol.* 336, 261–284.

Andersen, P. and B. Saltin (1985) Maximal perfusion of skeletal muscle in man. *J. Physiol.* 366, 233–249.

Anderson, C.R. and C.F. Stevens (1973) Voltage clamp analysis of acetylcholine produced end plate current fluctuations at frog neuromuscular junction. *J. Physiol.* 235, 655–691.

Arai, M., K. Otsu, D.H. MacLennan, and M. Periasamy (1992) Regulation of sarcoplasmic reticulum gene expression during cardiac and skeletal muscle development. *Am. J. Physiol.* 262, C614–C620.

Arora, K.K., C.R. Bilburn, and P.L. Pedersen (1993) Structure/function relationships in hexokinase. Site-directed mutational analyses and characterization of overexpressed fragments implicate different functions for the N- and C-terminal halves of the enzyme. *J. Biol. Chem.* 268, 18259–18266.

Arthur, P.G., M.C. Hogan, P.D. Wagner, and P.W. Hochachka (1992a) Understanding hypoxia effects on skeletal muscle metabolism. *J. Appl. Physiol.* 73, 737.

Arthur, P.G., T.G. West, R.W. Brill, P.M. Schulte, and P.W. Hochachka (1992b) Recovery metabolism of skipjack tuna white muscle: rapid and parallel changes of lactate and phosphocreatine after exercise. *Can. J. Zool.* 70, 1230–1239.

Atkinson, D.E. (1990) *Control of Metabolic Processes* (Ed. Cornish-Bowden, A. and M.L. Cardenas), Plenum Press, New York, pp. 11–27.

Atkinson, D.E. and M.N. Camien (1982) The role of urea synthesis in the removal of bicarbonate and the regulation of blood pH. *Curr. Top. Cell. Reg.*, 21, 261–302.

Balaban R.S., H.L. Kautor, L.A. Katz, R.W. Briggs (1990) Relation between work and phosphate metabolites in the *in vivo* paced mammalian heart. *Science* 232, 1121–1123.

Balaban, R.S. (1990) Regulation of oxidative phosphorylation in the mammalian cell. *Am. J. Physiol.* 258, C377–C389.

Baldwin, J. (1988) Adaptation, constraint, and convergence among vertebrate LDH isozymens: an evoluntionary perspective. *Can. J. Zool.* 66, 1011–1014.

Baldwin, J. and P.W. Hochachka (1985) A glycolytic paradox in *Limaria* muscle. *Mol. Physiol.* 7, 29–40.

Barchi, R.L. (1993) Ion channels and disorders of excitation in skeletal muscle. *Cur. Opinion Neurol. Neurosug.* 6, 40–47.

Bashan, N., E. Burdett, H.S. Hundal, and A. Klip (1992) Regulation of glucose transport and GLUT1 glucose transporter expression by O_2 in muscle cells in culture. *Am. J. Physiol.* 262, C682–C690.

Bessman, S.P. and P.J. Geiger (1981) Transport of energy in muscle: the phosphorylcreatine shuttle. *Science,* 211, 448–452.

Betts, D.F. and D.K. Srivastava (1991) The rationalization of high enzyme concentrations in metabolic pathways such as glycolysis. *J. Theor. Biol.* 151, 155–167.

Birnbaumer, I., X. Wei, A. Neely, R. Olcese, A. Castellano, H. Kim, E. Perez-Reyes, and E. Stafani (1993) Studies on structure and function of voltage-gated Ca^{++} channels. *Int. Union Physiol. Sci. Abstr.* 147.4/0.

Blatz, A.L. and K.L. Magleby (1983) Single voltage-dependent chloride-selective channels of large conductance in cultured rat muscle. *Biophys. J.* 43, 237–241.

Block, B.A. (1991) Endothermy in fish. In *Biochemistry and Molecular Biology of Fishes* (Ed. Hochachka, P.W. and T.P. Mommsen), Elsevier, Amsterdam.

Block, B.A., J. R. Finnerty, A.F.R. Stewart, and J. Kidd (1993) Evolution of endothermy in fish: mapping physiological traits on a molecular phylogeny. *Science* 260, 210–214.

Blum, H., S. Nioka, R.G. Johnson, Jr. (1990) Activation of the $Na^{+}K^{+}$ ATPase in *Narcine brasiliensis. Proc. Natl. Acad. Sci. U.S.A* 87, 1247–1251.

Blum, H., J.A. Balschi, and R.G. Johnson, Jr. (1991) Coupled *in vivo* activity of the membrane band $Na^{+}K^{+}$ ATPase in resting and stimulated electric organ of the electric fish *Narcine brasiliensis. J. Biol. Chem.* 266, 10254–10259.

Boiteux, A. and B. Hess (1981) Design of glycolysis. *Phil. Trans. R. Soc. Lond.* B293, 5–22.

Booth, F.W. and B.S. Tseng (1993) Olympic goal: molecular and cellular approaches to understanding muscle adaptation. *News Physiol. Sci.* 8, 165–169.

Boulter, J., K. Evans, D. Goldman, G. Martin, D. Teco, S. Heinmann, and J. Patrick (1986) Isolation of a cDNA clone coding for a possible neural nicotinic acetylcholine receptor α-subunit. *Nature* 319, 368–374.

Bowlus, R.D. and G.N. Somero (1979) Solute compatibility with enzyme function and structure: rationales for the selection of somotic agents and end products of anaerobic metabolism in marine invertebrates. *J. Exp. Zool.* 208, 37–152.

Brandl, C.J., N.M. Green, B. Korczak, and D.H. MacLennan (1986) Two Ca^{++} ATPase genes: Homologies and mechanistic implications of deduced amino acid sequences. *Cell* 44, 597–607.

Bremer, J. and H. Osmundsen (1984) Fatty acid oxidation and its regulation. In *Fatty Acid Metabolism and Its Regulation* (Ed. Numa, S.), Elsevier, Amsterdam, pp. 113–154.

Brooks, G.A. (1986) Lactate production under fully aerobic conditions: the lactate shuttle during rest and exercise. *Fed. Proc.* 45, 2924–2929.

Brooks, G.A., D.M. Donovan, and T.P. White (1984) Estimation of anaerobic energy production and efficiency in rats during exercise. *J. Appl. Physiol.* 56, 20–525.

Brown, A.M., K. Morimoto, Y. Tsuda, and D.L. Wilson (1981) Calcium current-dependent and voltage-dependent inactivation of calcium channels in *Helix aspersa. J. Physiol.* 320, 193–218.

Burke, E.R., F. Cerny, D. Costill, and W. Fink (1977) Characteristics of skeletal muscle in competitive cyclists. *Med. Sci. Sports* 9, 109–112.

Carilli, C.T., M. Berne, L.C. Cantley, and G.T. Haupert, Jr. (1985) Hypothalamic factor inhibits the (Na,K)ATPase from the extracellular surface. *J. Biol. Chem.* 260, 1027–1031.

Casey, T.M., C.P. Ellington, and J.M. Gabriel (1992) Allometric scaling of muscle performance and metabolism: insects. *Adv. Biosci.* 84, 152–162.

Catterall, W.A. (1986) The molecular basis of neuronal excitability. *Science* 223, 653–661.

Catterall, W.A. (1988) Ca^{++} channel structure and function. *Science* 242, 50–61.

Catterall, W.A. (1991) Excitation-contraction coupling in vertebrate skeletal muscle: a tale of two calcium channels. *Cell* 64, 871–874.

Catterall, W.A. (1992) Cellular and molecular biology of voltage-gated sodium chanels. *Physiol. Rev.* 72, 515–548.

Chance, B., S. Eleff, J.S. Leigh, Jr., D. Sokolow, and A. Sapega (1981) Mitochondrial regulation of phosphocreatine/inorganic phosphate ratios in exercising human muscle: a gated 31P NMR study. *Proc. Natl. Acad. Sci. U.S.A.* 78, 6714–6718.

Chance. B., J.S. Leigh, Jr., J. Kent, and K. McCully (1986) Metabolic control principles and 31P NMR. *Fed. Proc.* 45, 2915–2920.

Charlemagne, D., J.-M. Maixent, M. Preteseille, L.G. Lelievre (1986) Ouabain binding sites and (Na^+, K^+)-ATPase activity in rat cardiac hypertrophy. *J. Biol. Chem.* 261, 185–189.

Chi, M.M.Y., C.S. Hintz, E.F. Coyle, W.H. Martin III, J.L. Ivy, P.M. Nemeth, J.O. Holloszy, and O.H. Lowry (1983) Effects of detraining on enzymes of energy metabolism in individual human muscle fibers. *Am. J. Physiol.* 244, C276–C287.

Clausen, T. (1986) Regulation of active Na^+-K^+ transport in skeletal muscle. *Physiol. Rev.* 66, 542–590.

Clausen, T. (1990) Significance of Na^+-K^+ pump regulation in skeletal muscle. *NIPS* 5, 148.

Cohen S.A. and R.L. Barchi (1993) Voltage-dependent sodium channels. *Int. Rev. Cytol.* 137C, 55–103.

Cohen, C.J., B.P. Bean, T.J. Colatsky, and R.W. Tsien (1981) Tetrodotoxin block of sodium channels in rabbit Purkinje fibers: interactions between toxin binding and channel gating. *J. Gen. Physiol.* 78, 383–411.

Collicutt, J.M. and P.W. Hochachka (1977) The anaerobic oyster heart: coupling of glucose and aspartate fermentation. *J. Comp. Physiol.* 115, 147–157.

Connett, R.J. (1988) Analysis of metabolic control: new insights using scaled creatine kinase model. *Am. J. Physiol.* 254, R949–959.

Connett, R.J. and C.R. Honig, (1989) Regulation of $\dot{V}O_2max$. Do current biochemical hypotheses fit *in vivo* data? *Am. J. Physiol.* 256, R898–R906.

Connett, R.J., C.R. Honig, T.E.J. Gayeski, and G.A. Brooks (1990) Defining hypoxia. *J. Appl. Physiol.* 63, 833–842.

Connett, R.J., T.E. Gayeski, and C.R. Honig (1985) Energy sources in fully aerobic rest-work transitions: a new role for glycolysis. *Am. J. Physiol.* 248, H922–H929.

Cooke, R. and E. Pate (1990) *Biochemistry of Exercise* (Ed. Taylor, A.W.), Human Kinetics Books, Champaign, IL. pp. 59–72.

Dann, L.G. and H.G. Britton (1978) Kinetics and mechanism of action of muscle pyruvate kinase. *Biochem. J.* 169, 39–54.

Davies, K.J.A., L. Packer, and G.A. Brooks (1981) Biochemical adaptation of mitochondria, muscle, and whole-animal respiration to endurance training. *Arch. Biochem. Biophys.* 209, 539–554.

Dawson, M.J., D.G. Gadian, and D.R. Wilkie (1978) Muscular fatigue investigated by phosphorus nuclear magnetic resonance. *Nature (Lond.)* 274, 861–866.

De Zwaan, A. (1983) Carbohydrate catabolism in bivalves. In *The Mollusca: Metabolic Biochemistry and Molecular Biomechanics,* Vol. 1 (Ed. Hochachka, P.W.), Academic Press, New York pp. 137–175.

Diamond J.M. and K.A. Hammond (1992) Intestinal determinants of muscle performance. *Hypoxia and Mountain Medicine* (Ed. Sutton, J.R., G. Coates, and C.S. Houston), Pergamon Press, Oxford, pp. 163–170.

Diamond J.M. and K.A. Hammond (1992) The matches, achieved by natural selection, between biological capacities and thier natural loads. *Experientia* 48, 551–556.

Dionne, V.E. (1981) The kinetics of slow muscle acetylcholine-operated channels in the garter snake. *J. Physiol.* 310, 159–190.

Dix, J.A. and A.S. Verkman (1990) Mapping of fluorescence anistropy in living cells by ratio imaging. Application to cytoplasmic viscosity. *Biophys. J.* 57, 231–240.

Dobson, G.P. and P.W. Hochachka (1987) Role of glycolysis in adenylate depletion and repletion during work and recovery in teleost white muscle. *J. Exp. Biol.* 129, 125–140.

Dobson, G.P., W.S. Parkhouse, J.M. Weber, E. Stuttard, J. Harman, D.H. Snow, and P.W. Hochachka (1988) Metabolic changes in skeletal muscle and blood in greyhounds during 800 m track sprint. *Am. J. Physiol.* 255, R513–R519.

Drake, A. (1985) Substrate utilization in the myocardium. *Basic Res. Cardiol.* 19, 1–11.

Driedzic, W.R. and P.W. Hochachka (1976) Control of energy metabolism in fish white muscle. *Am. J. Physiol.* 230, 579–582.

Dudley, G.A., P.C. Tullson, and R.L. Terjung (1987) Influence of mitochondrial content on the sensitivity of respiratory control. *J. Biol. Chem.* 262, 9109–9114.

Ebashi, S. (1983) Regulation of muscle contraction. *Cell Muscle Motility* 3, 79–87.

Ebashi, S. (1991) Excitation-contraction coupling and the mechanism of muscle contraction. *Ann. Rev. Physiol.* 53, 1–16.

Eckardt, K.U., C.W. Pugh, P.J. Ratcliffe, and A. Kurtz (1993) Oxygen dependent expression of the erythropoitin gene in rat hepatocytes in vitro. *Pflugers Arch.* 423, 356–364.

Ellington, W.R. (1989) Phosphocreatine represents a thermodynamic and functional improvement over other muscle phosphagens. *J. Exp. Biol.* 143, 177–194.

England, W.R. and J. Baldwin (1983) Anaerobic energy metabolism in the tail musculature of the Australian yappy, *Cherax destructor* role of phosphagens and anaerobic glycolysis during escape behaviour. *Physiol. Zool.* 56, 614–622.

Eppenberger-Eberhardt, M., I. Riesinger, M. Messerli, P. Schwarb, M. Muller, H.M. Eppenberger, and T. Wallimann (1991) Adult rat cardiomyocytes cultured in creatine-deficient medium display large mitochondria with paracrystalline inclusions enriched in creatine kinase. *J. Cell Biol.* 113, 289–302.

Fertuck, H.C. and M.M. Salpeter (1976) Quantitation of junctional and extrajunctional acetylcholine receptors by electron microscope autoradiography after ^{125}I-α-bungarotoxin binding at mouse neuromuscular junctions. *J. Cell Biol.* 69, 144–158.

Fong, P., P.R. Turner, W.F. Denetclaw, and R.A. Steinhardt (1990) Increased activity of calcium leak channels in myotubes of duchenne human and *mdx* mouse origin. *Science* 250, 673–676.

From, A.H.L., S.D. Zimmer, S.P. Michurski, P. Mohanakrishnan, V.K. Ulstad, W.J. Thomas, and K. Ugurbil (1990) Regulation of oxidative phosphorylation in the intact cell. *Biochemistry* 29, 3733–3743.

Funk, C.I., A. Clark, Jr., and R.J. Connett, (1990) A simple model of metabolism: applications to work transitions in muscle. *Am. J. Physiol.* 258, C995–C1005.

Gade, G. (1986) Purification and properties of tauropine dehydrogenase from the shell adductor muscle of the Ormer, *Haliotis lamellosa*. *Eur. J. Biochem.* 160, 311–318.

Gadian, D.G., G.K. Radda, T.K. Brown, E.M. Chance, M.J. Dawson, and D.R. Wilkie (1981) The activity of creatine kinase in frog skeletal muscle studied by saturation-transfer nuclear magnetic resonance. *Biochem. J.* 194, 215–228.

Gayeski, T.E.J. and C.R. Honig (1986) O_2 gradients from sarcolemma to cell interior in red muscle at maximal VO_2. *Am. J. Physiol.* 251, 789–H799.

Geers, C., P. Wetzel, and G. Gros (1991) Is carbonic anhydrase required for contraction of skeletal muscle? *News Physiol. Sci.* 6, 78–81.

Gelb, B.D, V. Adams, S.N. Jones, L.D. Griffin, G.R. MacGregor, and E.R.B. McCabe (1992) Targeting of hexokinase 1 to liver and hepatoma mitochondria. *Proc. Natl. Acad. Sci. U.S.A.* 89, 202–206.

Grabarek, Z., T. Tao, and J. Gergely (1992) Molecular mechanism of troponin-C function. *J. Muscle Res. Cell Motility* 13, 383–393.

Graham, T.E., J. Bangsbo, P.D. Gollnick, C. Juel, and B. Saltin (1990) *Am. J. Physiol.* 259, E170–E176.

Gros, G. and S.J. Dodgson (1988) Velocity of CO_2 exchange in muscle and liver. *Annu. Rev. Physiol.* 50, 669–694.

Guppy, M., W.C. Hulbert, and P.W. Hochachka (1979) Metabolic sources of heat and power in tuna muscles. II. Enzyme and metabolite profiles. *J. Exp. Biol.* 82, 303–320.

Hackney, D.D. and P.K. Clark (1985) Steady state kinetics at high enzyme concentration. The myosin MgATPase. *J. Biol. Chem.* 260, 5505–5510.

Harrison, S.M. and B.M. Bers (1990) Modifiction of temperature dependence of myofilament Ca sensitivity by troponin C replacement. *Am. J. Physiol.* 258: C282–C288.

Harrison, M.L., P. Rathinavelu, P. Arese, R.L. Geahlen, and P.S. Low (1991) Role of band 3 tyrosine phosphorylation in the regulation of erythrocyte glycolysis. *J. Biol. Chem.* 266, 4106–4111.

Hastings, K.E., E.A. Bucher, and C.P. Emerson, Jr. (1985) Generation of troponin T isoforms by alternative RNA splicing in avian skeletal muscle. Conserved and divergent features in birds and mammals. *J. Biol. Chem.* 260, 13699–13703.

Heilbrunn, L.V. (1956) *The Dynamics of Living Protoplasm.* Academic Press, New York, pp. 1–325.

Hill, A.V. (1950) The dimensions of animals and their muscular dynamics. *Sci. Prog. (Oxford)* 150, 209–230.

Hille, B. (1984) *Ionic Channels of Excitable Membranes.* Sinauer Assoc. Inc., Sunderland, MA, pp. 1–426.

Hille, B. (1992) *Ionic Channels of Excitable Membranes.* Sinauer Assoc. Inc., Sunderland, MA, pp. 1–607.

Hochachka, P.W. (1973) Metabolism. In *Comparative Animal Physiology* (Ed. Prosser, C.L.), W.B. Saunders, Philadelphia.

Hochachka, P.W. (1980) *Living Without Oxygen.* Harvard University Press, Cambridge, MA, pp. 1–181.

Hochachka, P.W. (1983) Protons and glucose metabolism in shock. *Adv. Shock Res.* 9, 49–65.

Hochachka, P.W. (1986) Balancing conflicting metabolic demands of exercise and diving. *Fed. Proc.* 45, 2948–2952.

Hochachka, P.W. (1987) Fuels and pathways as designed systems for support of muscle work. *J. Exp. Biol.* 115, 149–164.

Hochachka, P.W. (1987) Patterns of O_2 dependence of metabolism. *Adv. Exp. Med. Biol.* 222, 143–149.

Hochachka, P.W. (1988a) Metabolic, channel, and pump-coupled functions: constraints and compromises of coadaptation. *Can. J. Zool.* 66, 1015–1027.

Hochachka, P.W. (1988b) Patterns of O_2 dependence of metabolism. *Adv. Exp. Med. Biol.* 222, 143–149.

Hochachka, P.W. (1988c) The lactate paradox: analysis of underlying mechanisms. *Ann. Sports Med.* 4, 184–189.

Hochachka, P.W. and R.A. Foreman III (1993) Phocid and cetacean blueprints of muscle metabolism. *Can. J. Zool.* 71, 2089–2098.

Hochachka, P.W. and M. Guppy (1987) Metabolic arrest and the control of biological time. Harvard University Press, Cambridge, MA, pp. 1–237.
Hochachka, P.W. and G.O. Matheson (1992) Regulation of ATP turnover over broad dynamic muscle work ranges. *J. Appl. Physiol.* 73, 570.
Hochachka, P.W. and T.P. Mommsen (1983) Protons and anaerobiosis. *Science,* 219, 1391–1397.
Hochachka, P.W. and C. Monge (1994) *Hypoxia Defense Adaptations in Man* Harvard Univeristy Press, Boston, MA, in preparation.
Hochachka, P.W. and G.N. Somero (1984) *Biochemical Adaptation,* Princeton University Press, New Jersey, pp. 1–521.
Hochachka, P.W., M. Bianconcini, W.S. Parkhouse, and G.P. Dobson (1991) Role of actomyosin ATPase in metabolic regulation during intense exercise. *Proc. Natl. Acad. Sci. U.S.A.,* 88, 5764–5768.
Hochachka, P.W., G.P. Dobson, and T.P. Mommsen (1983a) Role of isozymes in metabolic regulation during exercise: insights from comparative studies. In *Isozymes—Current Topics in Biological and Medical Research,* Vol. 8, (Eds. Rattazzi, M.C., J.G. Scandalios, and G.S. Whitt), Alan R. Liss, New York, pp. 91–113.
Hochachka, P.W., B. Emmett and R.K. Suarez (1988) Limits and constraints in the scaling of oxidative and glycolic enzymes in homeotherms. *Can. J. Zool.* 66, 1128–1138.
Hochachka, P.W., T.P. Mommsen, J.H. Jones, and C.R. Taylor (1987) Substrate and O_2 fluxes during rest and exercise in a high altitude adapted animal, the Llama. *Am. J. Physiol.* 253, R298–R305.
Hochachka, P.W., W.B. Runciman, and R.V. Baudinette (1985) Why exercising tammar wallabies turn over lactate rapidly: implications for models of mammalian exercise metabolism. *Mol. Physiol.* 7, 7–28.
Hochachka, P.W., C. Stanley, G.O. Matheson, D.C. McKenzie, P.S. Allen, and W.S. Parkhouse (1991) Metabolic and work efficiencies during exercise in Andean natives. *J. Appl. Physiol.* 70, 1720–1730.
Hochachka, P.W., C. Stanley, D.C. McKenzie, A. Villena, and C. Monge (1992) Enzyme mechanisms for pyruvate-to-lactate flux attentuation: study of Sherpas, Quechuas, and hummingbirds. *Int. J. Sport Med.* 13, S119–123.
Hodgkin, A.L. and A.F. Huxley (1952a) The components of membrane conductance in the giant axon of *Loligo. J. Physiol.* 116, 473–496.
Hodgkin, A.L. and A.F. Huxley (1952b) The dual effect of membrane potential on sodium conductance in the giant axon of *Loligo. J. Physiol.* 116, 497–506.
Hoffer, J.A., N. Sugano, G.E. Loeb, W.B. Marks, M.J. O'Donovan, and C.A. Pratt (1987) Cat hindlimb motoneurons during locomotion. II. Normal activity patterns. *J. Neurophysiol.* 57, 530–553.
Hogan, M.C., P.G. Arthur, D.E. Bebout, P.W. Hochachka, and P.D. Wagner (1992) The role of O_2 in regulating tissue respiration in dog muscle working *in situ. J. Appl. Physiol.* 73, 728.
Holloszy, J.O. and E.F. Coyle (1984) Adaptations of skeletal muscle to endurance exercise and their metabolic consequences. *J. Appl. Physiol.* 56, 831–838.
Holloszy, J.O. and F.W. Booth (1976) Biochemical adaptations to endurance exercise in muscle. *Annu. Rev. Physiol.* 38, 273–291.
Horowitz, S.B. and D.S. Miller (1984) Solvent properties of ground substance studied by cryomicrodissection and intracellular reference phase techniques. *J. Cell Biol.* 99, 172s–179s.

Howald, H.M., G. Von Glutz, and R. Billeter (1978) Energy stores and substrate utilization in muscle during exercise. 3rd International Symposium. Symposia Specialists, Inc., pp. 75–86.

Hulbert, A. (1987) Thyroid hormones, membranes, and evolution of endothermy. In *Advances in Physiological Research* (Ed. MacLennan, H.), Plenum Press, New York, pp. 305–320.

Hurley, B.F., J.M. Hagberg, W.K. Allen, D.R. Seals, J.C. Young, R.W. Cuddihee, and J.O. Holloszy (1984) Effect of training on blood lactate levels during submaximal exercise. *J. Appl. Physiol.* 56, 1260–1264.

Huxley, H.E. (1985) The crossbridge mechanism of muscular contraction and its implications. *J. Exp. Biol.* 115, 17–30.

Issekutz, B., Jr., W.A.S. Shaw, and A.C. Issekutz (1976) Lactate metabolism in resting and exercising dogs. *J. Appl. Physiol.* 40, 312–319.

Jacobs, H.K. and S.A. Kuby (1980) Studies on muscular dystrophy. A comparison of the steady-state kinetics of the normal human ATP-creatine transphosphorylase isoenzymes (creatine kinases) with those from tissues of Duchenne muscular dystrophy. *J. Biol. Chem.* 255, 8477–8482.

Jacobus, W.E., R.W. Moreadith, and K. M. Vandegaer (1982) Mitochondrial respiratory control. Evidence against the regulation of respiration by extramitochondrial phosphorylation potentials or by [ATP]/[ADP] ratios. *J. Biol. Chem.* 257, 2397–2402.

Jones, N.L., G.J.F. Heigenhauser, A. Kuksis, C.G. Matsos, J.R. Sutton, and C.J. Toews (1980) Fat metabolism in heavy exercise. *Clin. Sci.* 59, 469–478.

Josephson, R.K. and D. Young (1985) A synchronous insect muscle with an operating frequency greater than 500 Hz. *J. Exp. Biol.* 118, 185–208.

Kanno, T., K. Sudo, I. Takeuchi, S. Kanda, N. Honda, Y. Nishimura, and K. Oyama (1980) Hereditary deficiency of lactate dehydrogenase M subunit. *Clin. Chim. Acta,* 108, 267–276.

Katz, B. and R. Miledi (1967) The release of acetylcholine from nerve endings by graded electric pulses. *Proc. R. Soc. Lond. B* 167, 23–38.

Katz, B. and R. Miledi (1972) The statistical nature of the acetylcholine potential and its molecular components. *J. Physiol.* 244, 665–699.

King, C.E., M.J. Malinyshyn, J.D. Mewburn, S.M. Cain, C.K. Chapler (1993) *FASEB J. Abstr.* 7 (4), 4407.

Klein, S.C., R.C. Haas, M.B. Perryman, J.J. Billadello, and A.W. Strauss (1991) Regulatory element analysis and structural characterization of the human sarcomeric mitochondrial creatine kinase gene. *J. Biol. Chem.* 266, 18058–18065.

Knowles, R.G. and S. Moncada (1993) Nitric oxide as a signal in blood vessels. *Trends Biochem. Sci.* 17, 399–402.

Kushmerick, M.J. (1985) Patterns of mammalian muscle energetics. *J. Exp. Biol.* 115, 165–177.

Kushmerick, M.J., R.A. Meyer, and T.R. Brown (1992) Regulation of oxygen consumption in fast- and slow-twitch muscle. *Am. J. Physiol.* 263, C598–C606.

Land, B.R., T.R. Podleski, E.E. Salpeter, and M.M. Salpeter (1977) Acetylcholine receptor distribution on myotubules in culture correlated to acetylcholine sensitivity. *J. Physiol.* 269, 156–176.

Leijendekker, W.J., C. van Hardenveld, and A.A.H. Kassenaar (1983) The influence of the thyroid state on energy turnover during tetanic stimulation in the fast twitch (mixed type) muscle of rats. *Metabolism* 32, 615–621.

Levitskii, D.O., T.S. Levchenko, V.A. Saks, V.G. Sharov, and V.N. Smirnov (1977) Functional coupling between Ca^{++}-ATPase and creatine phosphokinase in sarcoplasmic reticulum of myocardium. *Biokimiia* 42, 1389–1395.

Lindstrom, J. (1985) Immunobiology of myasthenia gravis, experimental autoimmune myasthenia gravis, and Lambert-Eaton syndrome. *Annu. Rev. Immunol.* 3, 109–131.

Llinas, R., M. Sugimori, D.E. Hillman, and B. Cherksey (1992) Distribution and functional significance of the P-type, voltage dependent Ca^{++} channels in the mammalian central nervous system. *Trends Neurosci.* 15, 351–355.

Lopez-Barneo, J., A.R. Benot, and J. Urena (1993) Oxygen sensing and the electrophysiology of arterial chemoreceptor cells. *News Physiol. Sci.* 8, 191–195.

Low, P.S., R.K. Geahlen, E. Mehler, M.L. Harrison (1990) Extracellular control of erythrocyte metabolism mediated by a cytoplasmic tyrosine kinase. *Biomed. Biochim. Acta* 49, S135–S140.

MacLennan, D.H. (1990) Molecular tools to elucidate problems in excitation-contraction coupling. *Biophys. J.* 58, 1355–1365.

MacLennan, D.H., C.J. Brandl, B. Korczak, and N.M. Green (1985) Amino-acid sequence of a Ca^{++} + Mg^{++}-dependent ATPase from rabbit muscle sarcoplasmic reticulum, deduced from its complementary DNA sequence. *Nature* 316, 696–700.

MacLennan, D.H., C.J. Brandl, B. Korczak, and N.M. Green (1986) Ca^{++} ATPases: contribution of molecular genetics to our understanding of structure and function. In *Proteins of Excitable Membranes* (Ed. Hille, B. and D. Fambrough), John Wiley & Sons, New York,. pp. 287–300.

Maekawa, M., K. Sudo, S.-L. Li Steven, and T. Kano (1991) Genotypic analysis of families with lactate dehydrogenase A (M) deficiency by selective DNA amplification. (1991) *Hum. Genetics* 88, 34–38.

Mandel, G. (1992) Tissue specific expression of voltage sensitive sodium channels. *J. Membr. Biol.* 125, 193–205.

Maruyama, K., D.M. Clarke, J. Fujii, T.W. Loo, and D.H. MacLennan (1990) Expression and mutation of Ca^{++} ATPases of the sarcoplasmic reticulum. *Cell Motility Cytoskel.* 14, 26–34.

Mastro, A.M., M.A. Babich, W.D. Taylor, and A.D. Keith (1984) Diffusion of a small molecule in the cytoplasm of mammalian cells. *Proc. Natl. Acad. Sci. U.S.A.* 81, 3414–3418.

Matheson, G.O., P.S. Allen, D.C. Ellinger, C.C. Hanstock, D. Gheorghiu, D.C. McKenzie, C. Stanley, W.S. Parkhouse, and P.W. Hochachka (1991) Skeletal muscle metabolism and work capacity: a ^{31}P-NMR study of Andean natives and lowlanders. *J. Appl. Physiol.*, 70, 1963–1976.

Mathieu-Costello, O., P.J. Agey, R.B. Logemann, R.W. Brill, and P.W. Hochachka (1992) Capillary-fiber geometrical relationships in tuna red muscle. *Can. J. Zool.* 70, 1218–1229.

Mathieu-Costello, O., R.K. Suarez, and P.W. Hochachka (1992b) Capillary-to-fiber geometry and mitochondrial density in hummingbird flight muscle. *Resp. Physiol.* 89, 113–132.

Matsuno-Yagi, A. and Y. Hatefi (1985) Studies on the mechanism of oxidative phosphorylation. Catalytic site cooperativity in ATP synthesis. *J. Biol. Chem.* 260, 14424–14427.

Mattera, R., B.J.R. Pitts, M.L. Entman, L. Birnbaumer (1986) Guanine nucleotide regulation of a mammalian myocardial muscarinic receptor system. Evidence for homo- and heterotropic cooperativity in ligand binding analyzed by computer-assisted curve fitting. *J. Biol. Chem.* 260, 7410–7421.

Matthews-Bellinger, J. and M.M. Salpeter (1978) Distribution of acetylcholine receptors at frog neuromuscular junctions with a discussion of some physiological implications. *J. Physiol.* 279, 197–213.

Maughan, R.J., C. Williams, D.M. Campbell, and D. Hepburn (1978) Fat and carbohydrate metabolism during low intensity exercise: effects of the availability of muscle glycogen. *J. Appl. Physiol.* 39, 7–16.

McArthur, M. D., C.C. Hanstock, A. Malan, L.C.H. Wang, and P.S. Allen (1990) Skeletal muscle pH dynamics during arousal from hibernation measured by 31P NMR spectroscopy. *J. Comp. Physiol.* 160, 339–348.

McCleskey, E.W., M.D. Womack, and L.A. Fieber (1993) Structural properties of voltage-dependent calcium channels. *Intl. Rev. Cytol.* 137C, 39–54.

McGilvery, R.W. (1975) "The use of fuels for muscular work." In *Metabolic Adaptation to Prolonged Physical Exercise* (Eds. Howald, H. and J. R. Poortmans), Birkhauser Verlag, Basel, pp. 12–30.

McGilvery, R.W. (1983) *Biochemistry, a Functional Approach.* W.B. Saunders, Philadelphia.

McKenzie, C.D., L. Goodman, G.O. Matheson, B. Davidson, C.C. Nath, W.S. Parkhouse, P.S. Allen, and P.W. Hochachka (1991) Cardiovascular adaptations in Andean natives after 6 weeks of exposure to sea level. *J. Appl. Physiol.,* 70, 2650–2655.

Means, A.R., J.S. Tash, and J.G. Chafouleas (1982) Physiological implications of the presence, distribution, and regulation of calmodulin in eukaryotic cells. *Physiol. Rev.* 62, 1–39.

Meissner, G. (1984) Adenine nucleotide stimulation of Ca^{++}-induced Ca^{++} release in sarcoplasmic reticulum. *J. Biol. Chem.* 259, 2365–2374.

Mercer, R.W. (1993) Structure of the Na, K-ATPase. *Intl. Rev. Cytol.* 137C, 139–168.

Methfessel, C. and G. Boheim (1982) The gating of single calcium-dependent potassium channels is described by an activation/blockade mechanism. *Biophys. Struct. Mech.* 9, 35–60.

Meyer, R.A., H.L. Sweeny, and J. J. Kushmerick (1984) A simple analysis of the "phosphocreatine shuttle". *Am. J. Physiol.* 246, C365–C377.

Miller, C. (1991) 1990: Annus Mirabilis of potassium channel. *Science* 252, 1092–1096.

Milligan, C.L. and C.M. Wood (1986) Tissue intracellular acid-base status and the fats of lactate after exhaustive exercise in the rainbow trout. *J. Exp. Biol.* 123, 23–144.

Mommsen, T.P. and P.W. Hochachka (1988) The purine nucleotide cycle as two temporally separated metabolic units: a study of trout muscle. *Metabolism* 37, 552–556.

Mommsen, T.P., C.J. French, and P.W. Hochachka (1980) Sites and patterns of protein and amino acid utilization during the spawning migration of salmon. *Can J. Zool.* 58, 785–1799.

Morii, H. and Y. Tonomura (1983) The gating behaviour of a channel for Ca^{++}-induced Ca^{++} release in fragmented sarcoplasmic reticulum. *J. Biochem.* 93, 1271–1285.

Moyes, C.D., O.A. Mathieu-Costello, R.W. Brill, and P.W. Hochachka (1992) Mitochondrial metabolism of cardiac and skeletal muscles from a fast and a slow fish. *Can J. Zool.* 70, 1246–1253.

Murray, J.M. and A. Weber (1974) The cooperative action of muscle proteins. *Sci. Am.* 230, 58–71.

Murrell, W., D. Crane, and C. Masters (1993) Ontogenic activities and interactions of the lactate dehydrogenase isozymes with cellular structures in the guinea pig. *Mech. Ageing Dev.* 69, 37–52.

Nioka, S., Z. Argov, G.P. Dobson, R.E. Forster, H.V. Subramanian, R.L. Veech, and B. Chance (1991) Substrate regulation of mitochondrial oxidative phosphorylation in hypercapnic rabbit muscle. *J. Appl. Physiol.* 72, 521–528.

O'Brien, J., G. Meissner, and B.A. Block (1993) The fastest contracting muscles of nonmammalian vertebrates express only one isoform of the ryanodine receptor. *Biophys. J.* 65, 2418–2427.

Ovadi, J. (1991) Physiological significance of metabolic channelling. *J. Theor. Biol.* 152, 1–22.

Pagliaro, L. (1993) Glycolysis revisited: a funny thing happened on the way to the Krebs cycle. *News Physiol. Sci.* 8, 219–223.

Paine, P.L. (1984) Diffusive and nondiffusive proteins in vivo. *J. Cell. Biol.* 99, 1881–1895.

Parkhouse, W.S., G.P. Dobson, and P.W. Hochachka (1988) Control of glycogenolysis in rainbow trout muscle during exercise. *Can. J. Zool.* 66, 345–351.

Patlak, J.B., K.A.F. Gration, and P.N.R. Usherwood (1979) Single glutamate-activated channels in locust muscle. *Nature* 278, 643–645.

Perry, S.V. (1985) Properties of muscle proteins: a comparative approach. *J. Exp. Biol.* 115, 31–42.

Pette, D. and Staron, R.S. (1993) The molecular diversity of mammalian muscle fibers. *News Physiol. Sci.* 8, 153–157.

Pollack, G.H. (1983) The cross-bridge theory. *Physiol. Rev.* 63, 1049–1113.

Post, R.L. (1989) Seeds of sodium, potassium ATPase. *Annu. Rev. Physiol.* 51, 1–15.

Pressley, T.A. (1992) Phylogenetic conservation of isoform-specific regions within alpha-subunit of Na^+ K^+ ATPase. *Am. J. Physiol.* 262, C743–C751.

Quistorff, B., L. Johansen, and K. Sahlin (1992) Absence of phosphocreatine resynthesis in human calf muscle during ischaemic recovery. *Biochem. J.* 291, 681–686.

Reiser, P.J., R.L. Moss, G.G. Quilian, and M.L. Greaser (1985) Shortening velocity and myosin heavy chains of developing rabbit muscle fibers. *J. Biol. Chem.* 260, 14403–14405.

Ritchie, J.M. and R.B. Rogart (1977) The binding of saxitoxin and tetradotoxin to excitable tissue. *Rev. Physiol. Biochem. Pharmacol.* 79, 1–50.

Robinson, J.B., Jr., L. Inman, B. Sumegi, and P.A. Srere (1987) Further characterization of the Krebs tricarboxylic acid cycle metabolon. *J. Biol. Chem.* 262, 1786–1790.

Rose, R.J., D.R. Hodgson, T.B. Kelso, L.J. McCutcheon, T.A. Reid, W.M. Bayly, and P.D. Gollnick, (1988) Maximum O_2 uptake, O_2 debt and deficit, and muscle metabolites in thoroughbred horses. *J. Appl. Physiol.* 64, 781–788.

Rosenberg, R.L., P. Hess, J.P. Reeves, H. Smilowitz, and R.W. Tsien (1986) Ca^{++} channels in planar lipid bilayers: insights into mechanisms of ion permeation and gating. *Science* 231, 1564–1566.

Rossi, A.M., H.M. Eppenberger, P. Volpe, R. Cotrufo, and T. Wallimann (1990) Muscle type MM creatine kinase is specifically bound to sarcoplasmic reticulum and can support Ca^{++} uptake and regulate local ATP/ADP ratios. *J. Biol. Chem.* 265, 5258–5266.

Rowan, A.N. and E.A. Newsholme (1979) *Biochem. J.* 178, 209–216.

Ruegg, J.C. (1986) *Calcium in Muscle Activation. A Comparative Approach,* Springer Verlag, Berlin, pp. 1–300.

Rumsey, W.L., C. Schlosser, E.M. Nuutinen, M. Robiollo, and D.F. Wilson (1990) Cellular energetics and the oxygen dependence of respiration in cardiac myocytes is dated from adult rat. *J. Biol. Chem.* 265, 15392–15402.

Sacktor, B. (1976) Biochemical adaptations for flight in the insects. *Biochem. Soc. Symp.* 41, 111–131.

Sacktor, B. and E. Hurlbut (1966) Regulation of metabolism in working muscle *in vivo.* II. Concentration of adenine nucleotide, arginine phosphate, and inorganic phosphate in insect flight muscle during flight. *J. Biol. Chem.* 241, 632–634.

Sacktor, B. and E. Worsmer-Shavit (1966) Regulation of metabolism in working muscles *in vivo:*concentration of glycolytic intermediates. *J. Biol. Chem.* 241, 626–633.

Sahlin, K., G. Palmskog, and E. Hultman (1978) Adenine nucleotide and IMP contents of the quadriceps muscle in man after exercise. *Pflugers Arch. Ges. Physiol.* 374, 193–198.

Saltin, B. (1985) Malleability of the system in overcoming limitations: functional elements. *J. Exp. Biol.* 115, 45–345.

Scalettar, B.A., J.R. Abney, and C.R. Hackenbrock (1991) Dynamics, structure, and function are coupled in the mitochodrial matix. *Proc. Natl. Acad. Sci. U.S.A.* 88, 8057–8061.

Schulte, P.M., C.D. Moyes, and P.W. Hochachka (1992) Integrating metabolic pathways in postexercise recovery of white muscle. *J. Exp. Biol.* 166, 181–196.

Schultz, G., W. Rosenthal, J. Hescheler, and W. Trautwein (1991) Role of G proteins in calcium channel modulation. *Annu. Rev. Physiol.* 52, 275–292.

Seeherman, H.J., C.R. Taylor, G.M.O. Maloiy, and R.B. Armstrong (1981) Design of the mammalian respiratory system. II. Measuring maximum aerobic capacity. *Respir. Physiol.* 44, 11–23.

Sellers, J.R. and B. Kachar, (1990) Polarity and velocity of sliding filaments: control of direction by actin and of speed by myosin. *Science* 249, 406–408.

Shoubridge, E.A. and G.K. Radda (1984) A ^{31}PNMR study of skeletal muscle metabolism in rats depleted of creatine with the analogue B-quanidinopropionic acid. *Biochem. Biophys. Acta* 805, 79–88.

Shoubridge, E.A., R.A.J. Challiss, D.J. Hayes, and G.K. Radda (1985) Biochemical adaptation in the skeletal muscle of rats depleted of creatine with the substrate analogue B-quanidinopropionic acid. *Biochem. J.* 232, 125–131.

Shull, G.E., A. Schwarts, and J.B. Lingrel (1985) Amino-acid sequence of the catalytic subunit of the (Na^+ + K^+)ATPase deduced from a complementary DNA. *Nature* 316, 691.

Simonides, W.S., G.C. van der Linden, and C. van Hardeveld (1990) Thyroid hormone differentially affects mRNA levels of Ca^{++} ATPase isozymes of SR in fast and slow skeletal muscle. *FEBS Lett.* 274, 73–76.

Singer, S.J. and G.L. Nicolson (1972) The fluid mosaic model of the structure of cell membranes. *Science* 175, 720–731.

Smith, J.B., L. Smith, and B.L. Higgins (1985) Temperature and nucleotide dependence of Ca^{++} release by myo-inositol 1,4,5-trisphosphate in cultured smooth muscle cells. *J. Biol. Chem.* 260, 14413–14416.

Snow, D.H., R.C. Harris, and S.P. Gash, (1985) Metabolic responses of equine muscle to intermittent maximal exercise. *J. Appl. Physiol.* 58, 1689–1697.

Sorrentino, V. and P. Volpe (1993) Ryanodine receptors: how many, where, and why? *Trends Pharmacol. Sci.* 14, 98–102.

Spitzer, N.C. (1979) Ion channels in development. *Annu. Rev. Neurosci.* 2, 363–397.

Srivastava, D.K. and S.A. Bernhard (1986a) Enzyme-enzyme interactions and the regulation of metabolic reaction pathways. *Curr. Top. Cell. Reg.* 28, 1–68.

Srivastava, D.K. and S.A. Bernhard (1986b) Metabolite transfer via enzyme-enzyme complexes. *Science* 234, 1080–1086.

Srivastava, D.K., P. Smolen, G.F. Betts, T. Fukushima, H.O. Spivey, and S.A. Bernhard (1989) Direct transfer of NADH between alpha-glycerol phosphate dehydrogenase and lactate dehydrogenase: fact or misinterpretation? *Proc. Natl. Acad. Sci. U.S.A.* 86, 6464–6468.

Stainsby, W.N. and H.G. Welch (1966) Lactate metabolism of contracting dog skeletal muscle *in situ*. *Am. J. Physiol.* 211, 177–183.

Storey, K.B. (1985) Metabolic biochemistry of insect flight. In *Circulation, Respiration, and Metabolism.* (Ed. Gilles, R.), Springer-Verlag, Berlin, pp. 193–207.

Storey, K.B. and J.M. Storey (1983) Carbohydrate metabolism in cephalopods. In *The Mollusca: Metabolic Biochemistry and Molecular Biomechanics,* Vol 1, Ed. Hochachka, P.W., Academic Press, New York pp. 91–136.

Stuhmer, W. and W. Almers (1982) Photobleaching through glass micropipettes: sodium channels without lateral mobility in the sarcolemma of frog skeletal muscle. *Proc. Natl. Acad. Sci. U.S.A.* 79, 946–950.

Suarez, R.K. and C.D. Moyes (1992) Mitochondrial respiration in locust flight muscles. *J. Exp. Zool.* 263; 351–355.

Suarez, R.K., G.S. Brown, and P.W. Hochachka (1986) Metabolic sources of energy for hummingbird flight muscle. *Am. J. Physiol.* 251, R537–R542.

Suarez, R.K., J.R.B. Lighton, C.D. Moyes, G.S. Brown, C.L. Gass, and P.W. Hochachka (1990) Fuel selection in rufous hummingbirds: ecological implications of metabolic biochemistry. *Proc. Natl. Acad. Sci. U.S.A.* 87, 9207–9210.

Suarez, R.K., J.R.B. Lighton, G.S. Brown and O.A. Mathieu-Costello (1991) Mitochondrial respiration in hummingbird flight muscle. *Proc. Natl. Acad. U.S.A.* 88, 4870–4873.

Sun, M.K. and D.J. Reis (1992) Evidence nitric oxide mediates the vasopressor response in sino-denervated rats. *Life Sci.* 50, 555–565.

Sutton, J.R., N.L. Jones, and C.J. Toews (1981) The effect of pH on muscle glycolysis during exercise. *Clin. Sci.* 61, 331–337.

Swandulla, D., E. Carbonne, and H.D. Lux (1991) Do calcium channel classifications account for neuronal calcium channel diversity? *Trends Neurosci.* 14, 46–51.

Tager, J.M., A.K. Groen, R.J.W. Wanders, J. Duszynksi, H.V. Westerhoff, and R.C. Vervoorn (1983) Control of mitochondrial respiration. *Biochem. Soc. Trans.* 11, 40–43.

Tank, D.W., R.L. Huganir, P. Greengard, and W.W. Webb (1983) Patch-recorded single-channel currents of the purified and reconstituted *Torpedo* acetylcholine receptor. *Proc. Natl. Acad. Sci. U.S.A.* 80, 5129–5133.

Taylor, C.R., G.M.O. Maloiy, E.R. Weibel, V.A. Langman, J.M.Z. Kamau, H.J. Seeherman, and N.C. Heglund (1981) Design of the mammalian respiratory system. III. Scaling maximum aerobic capacity to body mass: wild and domestic mammals. *Respir. Physiol.* 44, 25–37.

Termin, A. and D. Pette (1991) Myosin heavy-chain based isomyosins in developing, adult fast-twitch and slow-twitch muscles. *Eur. J. Biochem.* 195, 577–584.

Thurman, R.G., Y. Nakagawa, T. Matsumura, J.J. Lemasters, U.K. Misra, and F.C. Kauffman (1993) Regulation of oxygen uptake in oxygen-rich periportal and oxygen-poor pericentral regions of the liver lobule by oxygen tension. In *Surviving Hypoxia—Mechanisms of Control and Adaptation* (Eds. Hochachka, P.W. et al.), CRC Press, Boca Raton, FL, pp. 329–340.

Toda, N. and T. Okamura (1992) Regulation by nitroxidergic nerve of arterial tone. *News Physiol. Sci.* 7, 148–152.

Trautwein, W. and J. Hescheler (1991) Regulation of cardiac L-type calcium current by phosphorylation and G proteins. *Annu. Rev. Physiol.* 52–257–274.

Vaghy, P.L. (1979) Role of mitochondrial oxidative phosphorylation in the maintenance of intracellular pH. *J. Mol. Cell. Cardiol.* 11, 33–940.

Van den Thillart, G., A. van Waarde, H.J. Muller, C. Erkelens, A. Addink, and J. Lugtenburg (1989) Fish muscle energy metabolism measured by *in vivo* ^{31}P-NMR during anoxia and recovery. *Am. J. Physiol.* 256, R922–R928.

van Deursen, J., A. Heerschap, F. Oerlemans, W. Ruttenbeek, P. Jap, H. ter Laak, and B. Wieringa, (1993) Skeletal muscles of mice deficient in muscle creatine kinase lack burst activity. *Cell* 74, 621–631.

Van Waarde, A., G. van den Thillart, C. Erkelens, A. Addink, and J. Lugtenburg (1990) Functional coupling of glycolysis and phosphocreatine utilization in anoxic fish muscle. *J. Biol. Chem.* 266, 914–917.

Veerkamp, J.H. and H.T.B. Moerkerk (1986) Peroxisomal fatty acid oxidation in rat and human tissues. *Biochem. Biophys. Acta* 875, 301–310.

Wagemann, G., E. Zanolla, H.M. Eppenberger, and T. Wallimann (1992) In situ compartmentation of creatine kinase in intact sarcomeric muscle: the actomyosin overlap zone as a molecular seive. *J. Muscle Res. Cell Motility* 13, 420–435.

Wallimann, T., T. Schlosser, and H.M. Eppenberger (1984) Function of the M-line bound creatine kinase as intramyofibrillar ATP regenerator at the receiving end of the phosphorylcreatine shuttle in muscle. *J. Biol. Chem.* 259, 5238–5246.

Wallimann, T., M. Wyss, D. Brdiczka, K. Nicolay, and H.M. Eppenberger (1992) Intracellular compartmentation, structure, and function of creatine kinase isozymes in tissues with high and fluctuating energy demands: the 'phosphocreatine circuit' for cellular energy homeostasis. *Biochem. J.* 281, 21–40.

Walker, J.B. (1979) Creatine: biosynthesis, regulation, and function. *Adv. Enzymol.* 50, 177–242.

Walsh, P.J., C. Bedolla, and T.P. Mommsen (1987) Reexamination of metabolic potential in the toadfish sonic muscle. *J. Exp. Zool.* 241, 133–136.

Wan, K.K. and J. Lindstrom (1985) Nicotinic acetylcholine receptor. In *The Receptors,* Vol. 1, Academic Press, New York, pp. 377–430.

Watanabe, M., T. Nagamine, K. Sakimura, Y. Takahashi, and H. Kondo. Developmental study of the gene expression for alpha and gamma subunits of enolase in the rat brain by in situ hybridization histochemistry. *J. Comp. Neurol.* 327, 350–358.

Waterston, R.H. and G.R. Francis (1985) Genetic analysis of muscle development in *Caenorhabditis elegans. Trends Neurosci.* 270–276.

Weber, J.-M., W.S. Parkhouse, G.P. Dobson, J. C. Harman, D.H. Snow, and P.W. Hochachka (1987) Lactate kinetics in exercising thoroughbred horses: II. Regulation of metabolite turnover rate in plasma. *Am. J. Physiol.* 253, R896–R903.

Webster, K.A. and B.J. Murphy (1988) Regulation of tissue-specific glycolytic isozyme genes: coordinate response to oxygen availability in myogenic cells. *Can. J. Zool.* 66, 1046–1058.

Wegener, G., N.M. Bolas, and A.A.G. Thomas (1991) Locust flight metabolism studied in vivo with 31P NMR spectroscopy. *J. Comp. Physiol.* B. 161, 247–256.

West, J. (1986) Lactate during exercise at extreme altitudes. *Fed. Proc.* 45, 2953–2957.

Whalen, R.G. (1985) Myosin isoenzymes as molecular markers in muscle physiology. *J. Exp. Biol.* 115, 43–52.

White, R.L. and B.A. Wittenberg (1993) NADH fluorescence of isolated ventricular myocytes: effects of pacing, myoglobin, and oxygen supply. *Biophys. J.* 65, 196–204.

Wicker, U., K. Bucheler, F.N. Gellerich, M. Wagner, M. Kapischke, and D. Brdiczka (1993) Effect of macromolecules on the structure of the mitochondrial intermembrane space and the regulation of hexokinase. *Biochem. Biophys. Acta* 1142, 228–239.

Wilkie, D.R. (1986) Muscular fatigue: effects of hydrogenions and inorganic phosphate. *Fed. Proc.* 45, 2921–2923.

Wittenberg, B.A. and J.B. Wittenberg (1987) Myoglobin-mediated oxygen delivery to mitochondria of isolated cardiac myocytes. *Proc. Natl. Acad. Sci. U.S.A.* 84, 7503–7507.

Wittenberg, J.B. and B.A. Wittenberg (1990) Mechanism of cytoplasmic hemoglobin and myoglobin function. *Ann. Rev. Biophys. Biophys. Chem.* 19, 217–241.

Woledge, R.C., N.A. Curtin, and E. Homsher (1985) *Energetic Aspects of Muscle Contraction.* Academic Press, New York pp. 1–357.

Yanagida, T., Y. Harada, and A. Ishijima (1993) Nano-manipulation of actomyosin molecular motors in vitro: a new working principle. *Trends Biochem. Sci.* 18, 319–324.

Yates, L.D. and M.L. Greaser (1983) Quantitative determination of myosin and actin in rabbit skeletal muscle. *J. Mol. Biol.* 168, 123–141.

INDEX

A

Acetylcholine (ACh), 13, 15, 16, 35, 36, 41, 59
Acetylcholinesterase (AChE), 33, 37
AcetylCoA, 86
N-Acetylglucosamine, 17, 21
ACh, see Acetylcholine
AChE, see Acetylcholinesterase
AChR, see ACh receptors
ACh receptor (AChR), 15, 18, 34, 36
  channel density, adjustment in, 36
  complex, purified, 18
  isoforms, 19, 20
Actin
  binding site for, 45
  isoform, 120
  structure, 50, 51
Actomyosin, 46, 55
AcylCoA synthetases, 90
Adenosine triphosphate (ATP), 2
  binding site, 64
  channeled arrival of, 57
  demand, 136
  -generating mechanisms, analysis of, 69
  hydrolysis, 46, 64, 82
  levels, buffering of, 77
  production, inhibition of, 116
  regeneration, 99
  replenishment, 77, 85
  saturation curves, 53
  synthase, 71
  synthesis, 73
    capacities, 126
    rates, 118
  turnover, 9, 92, 97
    model of control of, 107
    pathway of, 98
    rates, 41, 74, 77, 101, 105, 108, 112, 115
  utilization, 2, 7
Adenylate, 104
  depletion, 89
  fluxes, 119
Adult fast myosin, 50
Aerobic metabolism, 87, 89
Alanopine, 69
Alcohol fermentations, 83
Amino acids, 71, 78
AMP deaminase, functions of, 89
Anaerobic glycolysis, 10
Anaerobic metabolism, 93
Andean natives, 103
APK, see Arginine phosphokinase
Arg, see Arginine
Arginine (Arg), 74
Arginine phosphokinase (APK), 74, 81
Aspartate, 71
ATP supply and demand, integrating, 95–118
  biceps of the greyhound, 102
  biceps and soleus of the laboratory cat, 99–101
  calf muscle of variably adapted humans, 103
  controlling the physical state of ICF, 116–118
  energy coupling in anaerobically-driven muscles, 96
  exogenous control of energy coupling, 108–109
  flight muscle of insects, 104–105
  gastrocnemius of laboratory rabbit, 101–102
  gastrocnemius of laboratory rat, 99
  gracilis and gastrocnemius of the laboratory dog, 101
  human muscle at maximum aerobic work rates, 97
  isoform definition of muscle machines, 49–53, 119–136
    brain heater organ, 127–130
    defining muscle fiber types, 119–121
    electroplax, 130
    fish white muscle, 124–125
    fixed nature of lactate paradox in Andean natives, 134
    glycolytic function in chronic hypobaric hypoxia, 132
    isoform basis for plasticity of muscle function, 134–135
    issue of emergent properties, 135–136
    nature's fastest oxidative muscles, 121–124
    rattler muscle of rattlesnake, 127
    role of LDH isozymes in chronic hypobaric hypoxia, 133–134

role of MDH and LDH in chronic hypobaric hypoxia, 132–133
sound-producing muscles, 125–127
why muscles specialize into few different fiber types, 130–132
leg muscle of the thoroughbred, 102–103
O_2 sensing, 115–116
oxygen sensing in regulation of ATP turnover, 109–115
pathway intermediates and latent enzyme concept, 105–108
quantifying energy coupling, 95–96
serving small muscle mass with large cardiac output, 103
setting the ATP demand, 97–99
thermally driven change in ATP turnover rates, 105
ATP, see Adenosine triphosphate
ATPase, 55, 98
active site, 45
catalytic mechanisms in, 49
maximal turnover number of, 64
reaction rates, 53
turnover rates, 56

B

Band 3 protein, 107–108
Bee, metabolic rate, 42
Brain heater organ, 127–128

C

Ca^{2+} ATPase, 34
Ca^{++} channels, 13, 27
in cardiac muscle, 29
densities, SR, 63
function, 16
openings, voltage regulation of, 35
presynaptic, 36
response, speed of, 35
ryanodine receptor, 25–29
sarcolemmal-based, 38
SR, 32
turnover number of, 35
voltage regulated, 28
Calsequestrin, 30, 119
Ca^{++} pump, 52, 59
Carbon fuel fluxes, 119
Carbonic anhydrase, 33, 88
Cardiac actin, 50
Cardiac muscle type, 4
Cardiac pump, 93
Ca^{++} release, 26, 27
Catalase, 33
Catalytic rates, 54
Catecholamine regulation, 67
Cat muscles, metabolic scope for, 101
Channel(s)
densities, 34, 35
ion-flux capacities of, 34
isoforms
catalytic functions of, 31
tissue-specific, 20
protein structure, 37
turnover numbers for, 33
Cloning, 49
Contractile force, 49
Contractile proteins, 30
Contraction, sliding filament model of, 43
Contraction-relaxation cycles, 83
CPK, see Creatine phosphokinase
Cr, see Creatine
Creatine (Cr), 74, 95
Creatine phosphokinase (CPK), 6, 8, 9, 66, 69, 106
buffering, 92
cytosolic isozyme of, 75
Creatine shuttle, 6–9, 55–57, 76
Cross-bridges, 43
Cytochrome oxidase, 90
Cytosolic malate dehydrogenases, 135
Cytosolic viscosity, 116

D

Density adjustments, 35
Depolarization, 21
Diffusion limits, 32, 53–58
Direct handoff mechanisms
advantages of, 55
multienzyme complexes as, 54
DNA, prehistoric, 1

E

EC coupling, see Excitation-contraction coupling
ECF, see Extracellular fluid
Electric fish, CPK function in, 95–118

Electric fish, modified muscle in, 130
Electron transfer system (ETS), 71, 86, 93, 94, 134
Electrophorus, 21
Electrophorus electricus, 17, 95–118
Embryonic myosin, 50
End plate, 41
  channels, 13, 19, 35
  densities, AChR, 36
  depolarization, 20
  membrane, 13
  regions, 20
Enzyme(s)
  kinetics, Michaelis-Menten model of, 32
  sequential complex of, 86
  turnover numbers for, 33

e_o (enzyme concentration)
  definition, 106
  role in regulation, 97–108
Epinephrine, 80
EPO, see Erythropoitin
Erythropoitin (EPO), 109, 115, 116
Escherichia coli, 18
ETS, see Electron transfer system
Exchangers, 73
Excitation-contraction (EC) coupling, 4, 11, 131, 136
Exercise sciences, 1
Extracellular fluid (ECF), 4, 63, 66, 88
Eye, ciliary muscles of, 37–38

F

Fast-swimming fishes, 127
Fast twitch
  fibers, 29, 63, 82, 87
  glycolytic (FG) muscles, 41, 69
  glycolytic oxidative (FOG) muscles, 41,100, 123
  muscle, 10, 30, 83, 84, 88, 96, 136
Fat metabolism, analyses of, 91
Fermentable fuel, 78, 82
Fermentation, 69
FG, see Fast-twitch glycolytic muscles
Fish sonic muscle, 38
Flux rates, 90
Flux regulation, 97
FOG, see Fast-twitch glycolytic oxidative muscles
Foot proteins, 25, 27, 30, 39, 119
Foot structures, 27
Framework, 1–12
  coadaptation and emergent properties, 11–12
  coadaptation of energy demand and energy supply pathways, 6–10
  coadaptation and metabolic isozymes, 10–11
  isoform machinery for speeding up information transfer, 4–5
  setting, 1–4
  troponin C isoforms in speeding up contractile function, 5–6
Free SR (fSR), 60
  freeze-fractured, 60
  purification, 63
Frog skeletal muscle, K^+ channels of, 24
fSR, see Free SR

G

Galactose, 17
GAPDH, see Glyceraldehyde-3-phosphate dehydrogenase
Gating kinetics, 24
Gene cloning studies, 62
Genetic analysis, 49
Giant barnacle, 117
Giant squid, 117
Globular actin monomers, 48
Glucose, 17
  catabolism, 8
  transporter, 33
Glucose-6-phosphate (G6P), 7
Glutamate-activated channels, 31
Glyceraldehyde-3-phosphate dehydrogenase (GAPDH), 107, 120
Glycogen, 82, 87, 122
  mobilizing enzyme, 10
  pathways, 83
Glycogen → lactate pathway, 79
Glycogenolysis, spring coil view of, 92
Glycolysis
  accelerator function of, 85
  anaerobic, 10
  control of red blood cell, 108
  pathway of, 69
  regulation of, 107
Glycolytic pathway, down-regulation of, 131
G6P, see Glucose-6-phosphate

H

Halobacterium rhodopsin, 34
Heavy meromyosin (HMM), 44
Helix-loop-helix motif, 5
Hexokinase (HK), 7, 8, 120
HIF, see Hypoxia-inducible factor
HK, see Hexokinase
HMM, see Heavy meromyosin
Human knee extensors, 42
Human leg muscles, 42
Hummingbird, 42
Hummingbird flight, 57
Hummingbird flight muscle, 121, 122, 133
Hydrogen peroxide, 33
Hypoxia, 110, 114
  chronic hypobaric, 132, 134
  -inducible factor (HIF), 109, 110, 114
  induction site, 109

I

ICF, see Intracellular fluid
Immunochemical specificity, 49
IMP, see Inosine monophosphate
Inosine monophosphate (IMP), 95, 96
Insect flight, 42, 57
Insect flight muscles, 104, 106
Insect sonic muscles, 38, 57
Insect synchronous flight muscles, 38
Intracellular fluid (ICF), 4
Ion
  channels, catalytic capacities of, 33
  fluxes, 32, 119
  homeostasis, 67
  pumps, ATP-dependent, 63
  selective channel, 14
  transmission, channels for, 35
Ischemia, 113, 114
Isoforms, 2, 5
Isoproteins, 2
Isozymes, 2

J

Junctional SR (jSR), 4, 60

K

K^+ channel
  inactivation, 25
  mammalian, 31
k_{cat}, definition, 106
3-Ketosteroidisomerase, 33
5-3-Ketosteroid isomerase, 33
Kinetic ceiling, 54
Knee extensor, 103
Krebs cycle, 72, 86, 87, 89, 109, 120

L

Lactate, 71, 91, 132
  concentrations, 96
  dehydrogenase (LDH), 10, 11, 78, 122, 135
  formation, 133
  oxidases, 10
  production, 134
Lambert-Eaton syndrome (LES), 35, 36
LDH, see Lactate dehydrogenase
LES, see Lambert-Eaton syndrome
Light meromyosin (LMM), 44
Lipid-free membrane, 19
LMM, see Light meromyosin
Longitudinal sarcoplasmic reticulum (LSR), 25
LSR, see Longitudinal sarcoplasmic reticulum

M

Malate dehydrogenase-citrate synthase (MDH-CS), 86
Mammalian muscles
  amino acids in, 71
  ATP turnover rates in, 41
  fast-twitch type, 42, 84
Mannose, 17
Marine invertebrates, aerobic metabolism in, 87
Marlin, 127
MDH-CS, see Malate dehydrogenase-citrate synthase
Membrane
  activation, 35
  depolarization, 52
  metabolic rates, 42–58, 95–118
Metabolite
  concentration, 95
  enzyme-enzyme direct transfer of, 54
MG, see Myasthenia gravis
Mitochondrial malate dehydrogenases, 135

Moment-to-moment mechanism, most common, 66
Monoclonal antibodies, 23
Motor nerve terminals, 31
Motor neuron
flow of information from, 15
impulses, 35
Muscle(s)
activation, 106
anaerobic metabolism in, 70
contraction, 44, 45
glycolysis
enzymes in, 83
stimulation of, 81
manipulation, 1
membrane, 13
metabolic rates, 42–58, 95–118
plasticity of, 135
properties of, 63
shortening, 43
type
isoform, 6
specificity of, 119
work, ATPase turnover rates during, 42
Muscle machines, energy demand of, 41–58
actomyosin ATPases in solution, 53–54
adaptable vs. conservative aspects of contractile components, 49–51
actin isoforms, 50–51
co-occurrence of specific contractile isoforms, 51
myosin isoforms, 49–50
tropomyosin isoforms, 51
troponin isoforms, 51
ATPase coupling with filament movement, 45–48
co-occurrence of contractile and regulatory protein isoforms, 52–53
contractile cycle as channeled reaction sequence, 55–56
diffusion limitation of enzyme function, 54–55
evidence of preferential access to and from actomyosin ATPase, 57
excitation-contraction coupling, 52
globular and filamentous forms of actin, 45
minimizing Ca^{++} diffusion-based limits, 57–58
role of actomyosin ATPase in adaptation of muscle function, 53
sarcomere, 42–43
sliding filament model of contraction, 43–44
taking myosin apart to identify functional domains, 44–45
three functions of myosin, 44
troponin and tropomyosin mediate Ca^{++} regulation of muscle contraction, 48–49
Muscle machines, supplying with energy, 69–94
anaerobic glycolysis, 77–85
anaerobic end products, 82–83
ATP yields of anaerobic pathways, 78
enzyme and isozyme function, 80–82
glycogen, 78
hormones and neurotransmitters, 78–80
upper glycolytic limits in muscle, 83–85
basic ATP-synthesizing pathways in muscle, 69–74
nature of effective phosphagens, 74–77
ancillary roles of phosphagens, 76–77
osmotic or ionic effects of phosphagen mobilization, 76
phosphagen buffer ATP content, 75
phosphagen end products, 76
phosphagen storage, 74
pros and cons of phosphagens, 77
utilizing phosphagen, 74–75
oxidative metabolism, 86–94
coordinating aerobic and glycolytic pathways, 93–94
nature of endogenous aerobic fuels, 87–89
nature of exogenous fuels of aerobic muscle metabolism, 89–93
utilizing ATP, 74
Myasthenia gravis (MG), 36
Myofibrillar protein, 50
Myofilament volume densities, 90
Myosin
activities, 44
ATPases, 135
concentration, 42
fast-type, 51
isoforms, functions of, 50

N

Na^+-Ca^{++} exchange system, 15
Na^+ channel, 34
densities, 37
inactivation of, 21
isoforms, 23

localization, 24
properties, 24
proteins, 22
vertebrate cardiac, 23
Na^+ flux rates, 37
Na^+K^+ ATPase, 34
densities, 68
inhibition of, 9
isoforms of, 66
Na^+K^+ pump, 63, 67
Na^+ pump requirement, 65
Neonatal myosin, 50
Nerve-to-muscle information systems, design of, 31–39
channel densities adjusted upward, 34–35
channels as efficient catalysts, 31–33
design criteria for Na^+ channel functions, 37–38
design criteria for postsynaptic signal transduction, 36–37
design criteria for presynaptic signaling processes, 35–36
design criteria for TT and SR Ca^{++} channels, 38–39
low channel densities, 33–34
overall design principles for nerve-to-muscle information flow systems, 39
Nerve-to-muscle signals, 13–30
acetylcholine, 15
ACh-induced depolarization, 16–17
Ca^{++} channel isoforms, 28–29
channels, 14–15
dependence of ACh release upon Ca^{++} channels, 15–16
end-plate channel isoforms, 19
excitation-contraction coupling, 25–27
excitation-contraction coupling in fast- and slow-twitch muscles, 29–30
facilitating role of calsequestrin, 30
how signals get there, 13–14
isoforms of delayed rectifier K^+ channels, 24–25
localization of end-plate channels, 19–20
localization of Na^+ channels, 24
muscle action potentials, 21
Na^+ channel isoforms, 22–24
structure of end-plate channels, 17–19
structure of Na^+ channels, 21–22
synaptic transmission time, 20
Neuromuscular junction
ACh release at, 36
organization of, 20
Neurotoxin binding sites, 22–23
Neurotransmitter receptors, 136
Neutron diffraction analysis, 61
Nitric oxide synthase (NOS), 116
NMRS, see Nuclear Magnetic Resonance Spectroscopy
NOS, see Nitric oxide synthase
Nuclear Magnetic Resonance Spectroscopy (NMRS), 9

O

Octopine, 69
Oligosaccharide chains, 14, 21
Oxygen detection sites, 115
Oxygen diffusion, 73, 114
Oxygen regulation, 109–117
Oxygen sensing, 109–116
Oxygen uptake, 42, 97–117

P

PalmitylCoA, 86
PArg, see Phosphoarginine
Parvalbumin, 58, 120
Pasteur effect, 132
PCr, see Phosphocreatine
PDH, see Pyruvate dehydrogenase
PFK, see Phosphofructokinase
PGK, see Phosphoglycerate kinase
Phosphagen, 104
as ATP source, 74
storage, 77
Phosphate metabolites, high energy, 105
Phosphoarginine (PArg), 74
Phosphocreatine (PCr), 9, 69, 74, 76
Phosphofructokinase (PFK), 80
Phosphoglycerate kinase (PGK), 81
Phospholamban, 63
Phosphorylation site, 65
PK, see Pyruvate kinase
PKA, see Protein kinase A
Plasma glucose, 91
Postsynaptic membrane, 13, 16, 19
Potential energy, 103
Precontraction state, return to, 59–68

Ca^{++} ATPase catalytic cycle, 60
Ca^{++} ATPase genes, 62
Ca^{++} ATPases, 62
Ca^{++} ATPase and sarcoplasmic reticulum structure, 59–60
coadaptation and design properties of SR Ca^{++} ATPases, 62–63
functional significance of Na^+ pump isoforms, 66
long-term Na^+ K^+ ATPase regulatory mechanisms, 67
magnitude of postexercise Na^+ K^+ imbalance, 66
minimal design criteria for Na^+ K^+ ATPase as ion pump, 65
model of Ca^{++} ATPase structure, 61
muscle Na^+ K^+ ATPase functional design considerations, 67–68
Na^+K^+ ATPase catalytic cycle, 64
Na^+K^+ ATPase and Na^+ pump, 63–64
Na^+ K^+ ATPase structure, 64–65
Na^+ pump isoforms based on α- and β-subunit polymorphism, 65
short-term Na^+K^+ ATPase regulatory mechanisms, 66–67
Presynaptic membrane, 13
Protein
chemistry, 28
icebergs, 24
isoforms, specificity of, 119
kinase A (PKA), 79, 80
kinase regulation, 63
macromolecules, 33
tyrosine kinase (PTK), 108
PTK, see Protein tyrosine kinase
Pump
densities, adaptability of, 34
turnover numbers for, 33
Pyruvate, 33, 71
activation of glycolytic flux to, 132
dehydrogenase (PDH), 56, 81
inhibition, 133
kinase (PK), 80, 81, 135
push, 134

Q

Q_{10}, 33, 105
Quechua Indians, 132, 133

R

Rapid-flow kinetics, 53
Rattlesnakes, rattler muscles of, 105, 127–129
Reference phase technique, 117
Resting metabolic rate (RMR), 103
Rhodopseudomonas viridis, 34
RMR, see Resting metabolic rate
Ryanodine, 27, 29
Ryanodine receptor, 25–29

S

Sarcolemma, excitation of, 48
Sarcomere length, 44
Sarcoplasmic reticulum (SR), 5, 13, 25, 26, 31, 121
Ca^{++} pump in, 59
membrane, amplification of, 39
volume densities, 126
Saturation kinetics, 32
Signal, reception of, 36
Signal transmission
from motor nerve to muscle, 20
from nerve to muscle, 14
system, 31
Single ion channel electrophysiology, 28
Single-motor force, 45
Site-directed mutagenesis, 5–6
Skeletal muscle, 4, 43
Slow myosin, 50
Slow oxidative fibers, 124
Slow-twitch
fibers, 29, 63, 135
muscle, 5, 30, 60
Smooth muscle, 50
Sound-producing muscles, 125, 126
Sports medicine, 1
Squirrels, hibernating, 105
SR, see Sarcoplasmic reticulum
Standard protein analysis, 49
Steroid, 33
Stop-flow kinetics, 53
Strombine, 69
STX, 21, 23
Subunit–subunit interactions, 29
Superfast muscles, 37
Swordfish, 127
Synaptic transmission, normal, 17

T

Tauropine, 69
TC, see Terminal cisternae
Temperature coefficient, 33, 105
Tension time integral (TTI), 101, 102
Terminal cisternae (TC), 25, 60
Terminal dehydrogenases, 69
Thoroughbred horses, 102
Tissue differentiation, 23
Tn, see Troponin
TnC, see Troponin C
Torpedo, 17, 18
Transmembrane channels, 31, 61
Transmembrane helices, hydrophobic, 62
Transmitter release, 115
Transporters, turnover numbers for, 33
Transverse tubules (TT), 13, 25–27, 31, 37, 127, 136
  membrane depolarization, 27
  excitation of, 48
  Na^+ channels along, 34
Triactin, 33
Triglycerides, 87
Tropomyosin isoforms, 5, 120
Troponin (Tn), 5, 48, 51
Troponin C (TnC), 51
TT, see Transverse tubules
TTI, see Tension time integral
TTX, 21, 23
Tuna white muscle, 96, 124
Turnover number, 31–33, 50
Turnover rates, 32

V

Valinomycin, 33
Vertebrate fast muscles, 38
Vertebrate slow muscles, 38
Viscosity, 116
VO_2, 42, 97–117
Voltage sensitivity, 29

X, Y, Z

X-ray diffraction analysis, 61
Z-band, 43